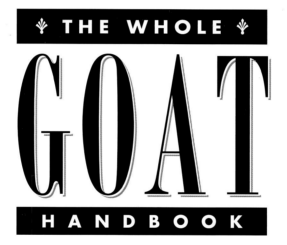

THE WHOLE
GOAT
HANDBOOK

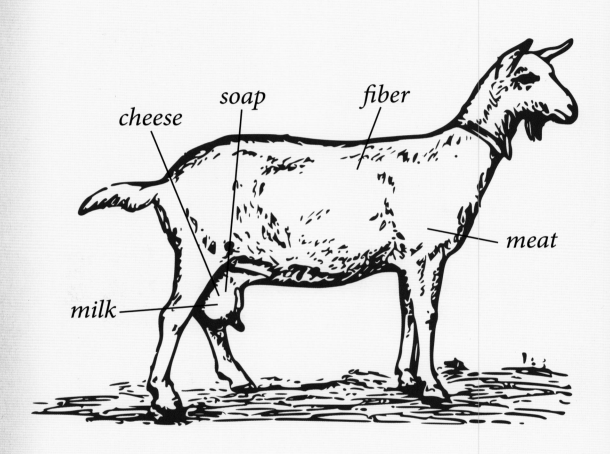

cheese

soap

fiber

milk

meat

THE WHOLE

GOAT

HANDBOOK

RECIPES, CHEESE, SOAP, CRAFTS & MORE

JANET HURST

Voyageur Press

Dedication

With love to my family: Charlie Hurst, Naomi Gottman, and James Bush.
"Through thick and through thin, throughout and through in, we're gonna get
through it together." With special memories of my late father, Albert Gottman,
who told me I should never buy a goat!

First published in 2013 by MBI Publishing Company LLC and Voyageur Press,
an imprint of MBI Publishing Company, 400 First Avenue North, Suite 300, Minneapolis, MN, 55401 USA

The information in this book is true and complete to the best of our knowledge. All recommendations are made
without any guarantee on the part of the author or Publisher, who also disclaim any liability incurred in connection
with the use of this data or specific details.

We recognize, further, that some words, model names, and designations mentioned herein are the property of the
trademark holder. We use them for identification purposes only. This is not an official publication.

MBI Publishing Company titles are also available at discounts in bulk quantity for industrial or sales-promotional
use. For details write to Special Sales Manager at MBI Publishing Company, 400 First Avenue North, Suite 300,
Minneapolis, MN, 55401 USA

ISBN: 978-0-7603-4236-7

Library of Congress Cataloging-in-Publication Data

Hurst, Janet.
  The whole goat handbook : recipes, cheese, soap, crafts & more / by Janet Hurst.
     p. cm.
  Includes bibliographical references and index.
  ISBN 978-0-7603-4236-7 (alk. paper)
  1. Goats. 2. Goat milk. 3. Goat cheese. 4. Goat meat. I. Title.
  SF383.H87 2013
  641.6'39--dc23
                    2012028074

Page 6: J. McPhail/Shutterstock.com
Back cover, right: Shutterstock.com

Design manager: Cindy Samargia Laun
Cover design: Brad Norr
Interior design: Brad Norr & Chris Fayers
Layout: Chris Fayers
Illustration: Mandy Kimlinger

Printed in China

Contents

Introduction

This book is written by a goat farmer for goat farmers. I am not a veterinarian, nor do I have formal training in veterinary science. I began my love affair with goats 25 years ago with the purchase of a single kid. From there I built a herd of 40 head, both angora and dairy goats. My adventures took me to employment on the largest dairy farm in Missouri, overseeing dairy goats and meat goats in numbers of 300 head (before kidding).

As you can imagine, experience quickly became my best teacher. A contract veterinarian set the protocols for routine care and management, which simplified the daily routines. However, animals are not known for good timing, and emergencies inevitably seemed to occur at night and on weekends or holidays, when I had the watch. I learned how to care for the herd by asking questions, shadowing the vet, then rolling up my sleeves with full knowledge that the health of the herd was in my hands. This book is a compilation of years of my own experience, combined with materials by goat experts from noted universities. This handbook will walk you through the process, from the selection of the animal, to an understanding of the breed, and on to the use of the products the animal will supply.

I hope you enjoy your journey with goats as much as I have. My life truly began the day I brought home my first one.

Goat Basics: Fact and Fiction

Goats. The mere mention of the word brings many images to mind. Children think of fairy tales complete with bridges and trolls; others share stories of goats eating labels from tin cans and tap dancing on top of their cars. There are many tales to tell, as goats remain one of the most creative and, yes, intelligent creatures on the planet. I know of no other animal that almost instantly brings a smile to the faces of those observing their antics. The kids hop, skip, and frolic their way through the pastures, making for hours of entertainment. Observing this carefree animal reminds us to stop and smell the roses, and maybe take a little taste of them, too!

✤ Goats across Time ✤

Goats are apparently as old as time itself—or at least certainly as old as the humans who have recorded their presence. Biblical references to goats are frequent, and the animal was often portrayed as a sacrificial animal, an image that has followed it throughout history. Cave dwellers depicted goats in primitive drawings; hieroglyphics also capture the goat as a part of everyday life in the Egyptian culture. Nomadic people found goats to be hardy, utilitarian beasts, and these animals have remained a part of their culture throughout history.

Studies show that goats are one of the earliest domesticated animals in Western Asia. The goat is thought to have descended from the Pasang or Grecian Ibex, a species of wild goat found in Asia Minor, Persia, and other nearby countries. A reference to the use of mohair from goats can be found in the Bible at the time of Moses when he told the children of Israel to bring white silk and goat's wool to weave altar cloths for the tabernacle.

Goats love a game of king of the mountain. Many goatkeepers add seesaws, barrels, or other equipment to provide goat entertainment. Providing a few things for goats to do keeps them content and less likely to wonder what is on the other side of the fence.

The first Swiss goat importations to the New World came from Switzerland. Records of early settlements in Virginia and New England indicate that milk goats were brought to the United States by Captain John Smith and Lord Delaware. Prior to 1904, very few Swiss goats occupied the United States. In that year, a consignment of 10 Saanens and 16 Toggenburg goats arrived, followed by more in the next 2 decades. According to the *Dairy Goat Production Guide* published by the University of Florida's Animal Sciences Department, descendants of these goats spread all over the country and provided the basis for the development and improvement of milk goats in this country.

Desert Goats

In my travels throughout Israel, I observed the Bedouin people, a nomadic tribe, traveling the desert with their entire encampment packed and carried by their camels. A herd of goats followed the caravan, their collar bells tinkling in the desert sun. Days later I saw the same tribe with their tents set up, the camels at rest, and the goats browsing on the scrub brush nearby. No fencing was used; the goats simply stayed by the camp, awaiting their master's hand for milking. When I inquired about the breed, I was told, "No breed. We call them desert goats." Research proved the animals to be that of Shami or Damascus goat family. This multipurpose animal produces milk and meat. The hides are tanned for leather and the animals are often trained to pull small carts or even a miniature plow.

Today, thoughts of goats often bring to mind an animal that appears to defy gravity on a mountainside. In truth, the animal's versatility is astounding as we look at its various habitats, ranging from desert to snow-covered peaks and from open prairie to farms. What other species is known for this level of adaptation? And as domesticated animals, their versatility continues to amaze. More than 300 breeds of goats are identified worldwide, with development of the various breeds for meat, dairy, fiber, goatskin, and the pet market. "Value added" is the buzzword in modern-day agriculture, and value can indeed be added to goat herds with the production of milk, cheese, and other dairy products; kidskin and pelts; yarn and clothing; and soap and lotions. Goats can also be trained as pack animals or to pull small carts.

Goats are employed as dairy animals all over the world. In fact, in the majority of European and underdeveloped countries, goat milk is consumed at a significantly higher rate that cow milk. Known for their utilitarian nature, goats can thrive and produce milk on a land base that would be unsuitable for dairy cattle. The smaller size of the goat also makes it "user friendly," and women and children can easily oversee the milking chores. In Africa, some goats are herded door to door and milk is bought and sold on order. The buyer presents an empty pot, and the goat herder milks the

Yes, LaManchas do have ears, but they're perfect, flat protrusions—nothing like those found on other breeds. This little girl has a look of surprise as she surveys the barnyard scene.

animal on the spot. Milk doesn't come any fresher than that! Cheesemaking is often a farmstead industry in these countries, and goats are the source for the milk that is processed into various types of soft and aged cheese.

Examples of goats adapting to meet the requirements of particular regions and climates of the world are quite evident. Goats in desert climates often sport lighter colored hair to deflect the heat of the day. Nubians originated in Africa, and Nubian bucks were crossed with goats of British origins to create the Anglo-Nubian Breed. The French breed, Alpine, is native to and quite comfortable in the cold mountainous regions of the Alps. The Toggenburg originated in Obertoggenburg, Switzerland; Saanens originated in Switzerland's Saanen Valley. Hearty Spanish meat goats and LaManchas originally came from Spain.

Goat breeds are distinguished by varying characteristics. Some have long-hair coats, such as angora and cashmere; the mountain goat is covered with lush winter fleece. Some are developed with specific milking qualities; others have a body structure most suitable for meat animals. Intentional breeding has strengthened desirable traits in individual breeds, and registered animals now number into the hundreds of thousands. Organizations such as the American Dairy Goat Society the International Goat Registry record the ancestry of the pedigreed animals.

⚜ Goat Myths and Realities ⚜

I once read that goats have "winning" personalities. In my years of experience, I have found this to be true. There is a particular pecking order and always a lead female goat. She will rule the herd, stand at the barn door, and let only her favorite herdmates enter. She will find this particularly entertaining on a rainy day. Her facial expression and body language speaks volumes.

One thing is certain, once you have been "owned" by a goat, there is little chance of recovery. Whether you are a backyard farmer with only a few animals or a herdsman of hundreds, something about this creature remains a fascinating challenge. To learn their habits, personality traits, and general nature is a lifelong study. Goats are thinkers. They will figure out ways to outsmart their handlers, unlock gates, and present themselves at the front door for supper.

Let's deal with some of the most common misinformation first. Goats do not eat tin cans. Whoever started this rumor must have witnessed an animal eating the label or licking the interior of the can for salt or sugar remnants. Goats do not eat "anything." In fact, they are quite particular about what they eat. They do not eat soiled hay, avoid certain plants that they know will not agree with them, prefer brush and foliage to grass, and drink only clean water. They need protection from bitter winter winds, and they hate to be in the rain. If they are to be at peak performance for milking, fiber growth, or poundage, then supplemental feed such as a total mixed ration (a.k.a. TMR, a formulated feed mix), routine parasite management, and a good water source are all factors that will determine the overall success of a goat project. Stocking rates vary with the quality of pasture or the amount of forage available. Ten to 12 goats per acre is a typical rate. Like any other living thing, goats have certain nutritional needs that must be met to maintain their health. What goes in comes out in the rate of gain and in the quality of the milk and fiber.

As a point of fact, though, goats will, however, chew on almost anything. Think of it this way: they are simply performing a taste test to see if a particular item is of interest to them. Clothes flapping on a clothesline are a tempting treat but are typically rejected after

sampling. Bright colored flowers, tree bark, and leaves are thoroughly enjoyed. A bit of car paint might come up missing, as the animals will lick the car for leftover bits of road salt.

The more you learn about goats, the more you'll find that anecdotal information on goats abounds. As my reputation of being the "goat lady" spread, many an older person approached me with tales of how goat milk saved his or her life as an infant. Goat milk was frequently used for delicate infants in generations past. Some pediatricians still prescribe goat milk today for cases of cow milk allergy and other conditions. The milk is a well-known remedy used to treat everything from ulcers to allergies. A common belief is that if you consume the milk of a goat that has ingested the poison ivy vine, you will become immune to that malady. The jury is still out on that subject, but it is an interesting notion. Goat milk also has been termed *naturally homogenized*, due to the fact that the fat globules stay suspended within the milk. This makes goat milk easier for some humans to digest when compared to cow milk. Goat milk is consumed all over the world, and in fact, it is the most frequently consumed milk worldwide.

Even our language gives a nod to the goat, and frequent colloquial references are made to them. Not all of the references are good, and they often reflect the goat's terrible reputation. Scapegoat (a reference to someone unjustly accused), old goat (a reference to someone usually cantankerous and well beyond the average lifespan), and many other common phrases are the result of the goat's influence upon our vernacular. However, if you truly have the opportunity to learn about goats and spend time with them, you will learn their habits and behaviors and still find that they

Goats are creatures of habit. They possess an internal clock and will go out to pasture and return at night on a particular schedule. This is fueled in part by a full udder and the awareness of milking time.

have a place in your farmstead plan, despite their colorful influence on our language.

As far as economics, grade (nonregistered) goats are not tremendously expensive to purchase or maintain. This viewpoint, while true, can be somewhat of a detriment to their ultimate survival as they are sometimes viewed as being disposable due to their relatively low investment and other prejudices against them. As usual, the best defense is information. Read all you can and learn to care for your own herd to the extent of your abilities. A great deal of research is available, and noted universities are a valuable resource. Some of those experts have contributed to this book. As goats gain in popularity, profes-

sionals are taking more interest, and it is only through research that we will learn more. Look for research-based information and the latest news and studies from universities and extension services (see the resource section, page 187, for additional information).

Keeping goats healthy takes the same dedication that any farmer would devote to any herd. Ask other goatkeepers to recommend a veterinarian who can help you with your goats; his or her medical knowledge will be invaluable at times. Educate yourself in common goat illnesses as well. Join a goat club and visit with 4-H leaders or other goat groups who speak the language of goats. You will soon learn to recognize symptoms, and treatment will become routine. Prevention is the best medicine a goat farmer can provide, and a good support group can help to demystify such skills as disbudding, hoof trimming,

Goats have the best personalities of any animal on the farm. This goat shares a laugh with her herd mates. I wonder what she knows that we don't?

Speaking Goat

Specific terms pertain to all aspects of farm life, including goatkeeping. Try these on for size.

- **Buck.** A male goat; also billie or, more formally, sires.
- **Buckling.** A male youngster.
- **Cabrito.** A cooking reference to a very young goat.
- **Capra aegagrus hircus.** The Latin name for goats. Goats are of the Caprine family.
- **Chevre.** Simple cheese made from goat milk.
- **Chevon.** French for goat.
- **Dam.** A female goat; also a doe or a nanny.
- **Disbudding.** The process of cauterizing the horn so that the growth will be inhibited.
- **Doeling.** A young female.
- **Fresh.** Term for a female goat in milk. A female must have an offspring to produce milk. Gestation is typically 145 to 158 days.
- **Horns.** Both males and females have them. Goats are born with tiny little buttons beneath the surface of their skin. These little buttons will form the horns.
- **Kids.** The offspring of goats. Twins are frequent; triplets and quads are common.
- **Polled.** A goat that does not grow horns. Some goats are naturally polled, which is a genetically induced condition.
- **Wether.** A castrated male.

Who knew there was a whole new language to learn? When I first started with goats and heard all those terms thrown about, I thought I would never learn all the terminology. Remember, there is no shame in asking about the meaning of a particular term, and as a general rule, goat people love to share.

castration, and other tasks that are a little nerve wracking to begin with. While no one relishes these tasks, it is all a part of keeping a healthy herd. Routine care such as worming at regular intervals, daily observation of the animals, vaccinations, and hoof care will save time, money, and headaches in the long run. A healthy goat should have bright and shining eyes, a cool dry nose with no discharge, and firm droppings. If you start to see a downcast spirit or other symptoms such as a dull coat, scours (diarrhea), or a general change in behavior, then start to look for the cause behind the change in the animal's presentation. Do not be penny wise and pound foolish: call in the professional if your instincts tell you to do so.

⚘ Meat, Milk, or Fiber ⚘

In the following chapters we will explore raising meat, milk, and fiber goats. All have similarities; in fact, they are more similar than different. However, their breeds lead them to their best use as a meat animal, a dairy animal, or one intended primarily for fiber. All three types of goats produce milk,

but only the dairy goat produces enough to feed her kids and beyond. The other two types primarily put forth energy into either gaining weight as animal protein or growing hair. The following traits are common among the meat, fiber, and dairy breeds:

- All goats are ruminants.
- Body structure is similar, with meat goats being the quickest to grow and put on body weight.
- Multiple births are common—and more so in the dairy breeds.
- Parasites and coccidosis are common threats.
- Goats are social, herd animals.
- They are browsers.
- Goats were nomadic, prior to domestication.

Indeed, if we were to view a goat without its outer coat, we would find many similarities among the meat, fiber, and dairy animals. The

Goats enjoy the company of others. When the decision is made to add goats to your farm, it is better to begin with two. One goat will be lonely and cry for companionship. In this case, two are better than one.

body structure, facial features, and musculo-skeletal structure are all very closely related. The meat goat has more fat and substance on the torso, the dairy goat develops a larger udder, and the fiber animal might be slightly shorter in stature; however, the differences are subtle. So if we look at the breeds as having more commonalities than differences for our study, we will understand the basic physiology of the goat breed as a whole.

Goats are ruminants. A ruminant digests its food by the transport of the vegetable matter through a series of stomachs. The large rumen, the rectulum, and the obomasum are the first three compartments; the fourth stomach is called the abomasum. The first three compartments of the digestive system do the

Carefree days on the goat farm. Spectacular Nubian ears are on display.

forework of digestion with the rumen serving as a fermenting compartment. Goats, like cows, will regurgitate larger fibrous materials and rechew them, as in chewing their cud.

Digestive enzymes are present to assist in the process, as are various digestive flora and microbial organisms. The rumen is the all-important precursor to true digestion. This compartment is where the breakdown of the food source begins. Digestive gasses assist in the assimilation of the foodstuff. If we put our ears down close to the goat's belly, we will hear all sorts of activity—lots of bubbling, growling, churning. This is the sign of an active rumen. If we listen and hear no activity, then we have problems. A common phrase for this condition would be to say that the goat has lost its rumen. In human terms, it would be a similar condition if we were to lose our digestive flora. Stomach maladies can occur as a result of overeating or gorging on a particularly rich substance. Notice that it is the same for humans and goats! The rec-

ommendation for fixing this ill is similar as well. Sometimes goats are given a dose of pink stomach meds or a prophylactic of digestive enzymes to restart the rumen.

Goats are herd animals and prefer to associate as a group. Until domestication, goats were nomadic. Problems with parasite management stem mainly from man's desire to confine them. If they were left to their own devices, they would travel each day, leaving their droppings behind as they move from place to place. However, when we fence animals and manure builds in an area, it is easy for goats to become overwhelmed with parasitic infestations. Unwittingly, parasites can be reingested and picked up from foliage when they are present in numbers representing overpopulation. Routine fecal samples are an important tool in overall herd health as well as manure management within confined areas. Samples can be collected and taken to a veterinarian for analysis, or you can learn to read them yourself with the aid of a microscope, a small lab, and some online research. Parasite management is one of the goatkeeper's ultimate challenges.

Al Johnson's Swedish Restaurant and Butik

When Al Johnson (www.aljohnsons.com) started his restaurant in Sister Bay, Wisconsin, many years ago, he could never have imagined that he was establishing a landmark, a destination, and a legend. Al Johnson's is an authentic Swedish, family-owned restaurant where you can find goats grazing the sod roof. While the traditional Scandinavian food is quite remarkable, the décor is inviting and the young women in Scandinavian dress serving limpa bread and Swedish meatballs all make for a wonderful dining experience, it is the goats on the roof that make this establishment known, quite literally, all over the world.

What began as a joke among friends—a goat as a birthday present from a mischief-making friend—soon became the hallmark of the restaurant. "Oscar" was installed on the sod roof of the restaurant, and soon others of his kind came to serve as the iconic figures of Al Johnson's Swedish Restaurant and Butik.

Al and his wife, Ingert, celebrated their Scandinavian and Swedish heritage throughout their years, sharing food with friends, neighbors, and customers of the uniquely themed restaurant. Since Al's passing, the restaurant remains in good hands, those of Rolf, Lars, and Annika Johnson, the couple's three children.

Changing with the times but keeping the best of the traditional garb, fare, and of course, the goats, the second generation of Johnsons recently brought technology to their operation. The addition of the Goat Cam allows daily, even hourly, checks on the sod roof lawnmowers! It is fun to go to the website and see what they are up to! The goats live a life of ease and enjoy their trip up their ramp each morning and back to the farm each night. Photographed by thousands, made popular in a children's book, and an Internet sensation, these animals have indeed gained worldwide fame.

RECIPES, CH

CRAF

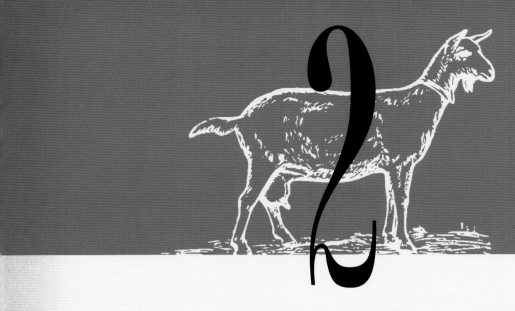

Goats for Milk

Dairy goats are one of the most versatile animals a small farmer can introduce to his or her homestead. They serve as utilitarian brush consumers, turning unwanted forage into nutrient rich milk. Goat milk contains a specific component, caprylic acid, which gives it that little bite, or tang. Some will find this slight aftertaste quite offensive; others will come to relish that little acidic note. For those who desire to make cheese, this bite will be an asset, adding a distinctive flavor that is unique to this particular milk.

Goat milk is excellent for drinking and has long been valued for its digestibility. At first sip, it may seem slightly thicker or more viscous to those who are not accustomed to it. The fat globules actually stay suspended in the milk. When compared to cow milk, the fat distribution is very different. If you have ever seen the cream rise to the top of cow milk of its own accord, it is simply a natural occurrence. Some breeds of goats have enough fat that this separation will also occur, but not to the same degree. Much of the fat (cream) stays integrated in the milk, to the point that it is somewhat difficult to separate. This is, however, what adds to the increased level of digestibility.

For those who desire to separate the cream from goat milk, for butter or to skim the milk, a cream separator is usually required. A little trick that also can be used is to pour the goat milk into a sun tea jar with a spigot. The lighter milk will drain off first, leaving the heavier cream behind. This method is not as effective as a cream separator, but it will work to capture quite a bit of the cream. Make sure to thoroughly sanitize the jar and spigot to avoid unwanted bacterial growth. A separator has a centrifugal action that throws the milk against the side of the bowl and literally beats the fat out of the milk. While an excellent healthy low-fat beverage, milk that has been separated is much more difficult

Goat milk is pure white and is the most widely consumed milk in the world. BestPhotoPlus/ Shutterstock.com

for making cheese. The central idea behind cheesemaking is to capture the milk solids, so when those have been greatly rearranged or broken in the separation process, it is quite difficult to make cheese.

Good goat milk begins with the diet of the animal. A poorly fed animal cannot produce rich milk. If you are keeping a goat primarily for milk, then supplemental grain will assist her in producing a good, rich product. While the forage will be the central part of the diet, the extra vitamins, minerals, and protein will aid the goat in the production of high quality milk. In the winter, if forage is not available, supplemental hay will be required.

❦ Purchasing a Dairy Goat ❦

Before you buy a goat, try milking a good one so you know what traits you are looking for in a quality animal. The udder should be well developed and well attached to the body. If it swings like a pendulum and looks like

Dairy Licenses

If you desire to sell milk, cheese, ice cream, or other dairy products, a state inspected facility may be required. Before you begin any commercial production, check with your state Department of Agriculture. This industry is heavily regulated and in most cases it is illegal to sell processed milk or dairy products without proper licensure.

it is supported by only a thin bit of muscle, then this is not your goat. Do not fall in love with just a pretty face! Though personality is important, if you are serious about milking, make sure the udder is firmly attached and symmetrical. Check the animal's coat. It

should be slick and shiny. If it isn't, the goat is probably full of parasites. Parasites can be dealt with, but again, it is best to begin with a healthy animal rather than to inherit problems to begin with.

The American Dairy Goat Association recognizes the following milking breeds: Alpine, Nubian, Oberhasli, Saanen, and LaMancha. Nubians are known for high butterfat content and Saanens for quantity of milk though lower butterfat content, which are considerations if you want to make cheese. The higher the fat content, the more cheese per gallon of milk. Some of the small breeds also are excellent for milk. Kinder goats, a cross between a Nubian and a pygmy, and Nigerian Dwarfs are smaller in stature, require less feed and produce an impressive amount of milk for their size.

⚜ Breeding ⚜

Goats typically produce milk for about 9 months after kidding. The amount of milk collected each day depends on multiple circumstances, including breed, diet, age of the animal, point of lactation, weather, and other factors. At peak production, one animal can give a gallon of milk (sometimes more) a day. Some give significantly less, and others even more. The production builds to a peak, remains there for a time, and then gradually tapers off. Many goatkeepers milk a goat for 10 months out of the year, rebreeding at 7 months into the milking cycle. The animals are milked for the first 2 months of their pregnancy, and then allowed to rest and to put their energies into the kids growing within their bodies. A goat's gestation cycle is about 5 months, and normally does have one pregnancy each year, which often results in multiple births. Twins, triplets, even quads are common.

The female goat goes through estrous, or heat cycles, often beginning in July. The most obvious sign of impending estrous is her behavior. She begins to rub on the fence, flick her tail, and bleat with great frequency. Her vulva becomes swollen, and she may have a clear discharge. At this point, the male becomes quite interested—and that is an understatement. If you have a breeding program in which you have a target date for your kids to be born (say to avoid the cold January conditions), then the buck must be kept away from the females. It is advisable to keep him at a totally separate farm or in a pen with a roof. A male goat interested in a female in heat will literally move mountains to mate. Sheds have been demolished, boulders have been displaced, and fences battered into splinters—all by a male goat in hot pursuit of his intended.

The male starts to perform a ritual to entice his lady friends. The goat equivalent of Aqua Velva is urine. The male urinates on his beard and forelegs; the greasy, musky spray is quite pleasing to the females of the herd. (If you come in contact with the male, be careful: this scent is hard to remove from your skin or clothing. Peppermint soap is the best cure.) If you are in doubt as to whether your does are indeed in heat, rub down the buck with a rag and place the rag in a jar. Then go to your does and open the jar. They will practically swoon if they are in heat. If they aren't, they will be oblivious to the scent.

Decide by your calendar when you want the kids to be born. March is a good time if winters are hard in your area. If March is the goal for kidding, then allow the buck to mate with the does in late October or early November. One buck can breed many does in a single night. Always remember, the animals may have other ideas about their breeding schedule. Just because you have a breeding plan does not mean things will go as scheduled. I've heard many stories of the buck escaping his pen and breeding an entire herd in one night. Of course, this was not according to the farmer's scheduled plan. But goats will be goats. The moral of this story? Give serious

Pygmies are short, stocky animals. When a female pygmy is bred with a Nubian male, the result is a Kinder goat. Kinders are known for their milk production levels despite their reduced size.

Saanens, which originated in Switzerland, are known for the large quantities of milk they produce and their low butterfat content. WilleeCole/ Shutterstock.com

thought to your buck pen. Do you need to keep a buck on the farm at all? Some people will rent them out. Artificial insemination is also a possibility.

A note about introducing a buck to a milking doe: Often people will say that goat milk has an off flavor and that is why a lot of folks don't enjoy goat milk. When the buck is in with the doe, expect that off flavor to be present in the milk. Limit the amount of time the buck is with the doe. The longer he is there, the more she will pick up his scent. Even with limited exposure, the milk may very well have a buckish taint. This problem is not as great in a large herd, as the taint will be somewhat diluted; however, in a small herd, it will be quite noticeable. This milk can be frozen and fed back to kids, fed to chickens or pigs, or used to make soap (see the soap-making section, Chapter 8). This is a good way to use milk that has the odor or isn't quite suitable for drinking or cheesemaking. Whatever you do, don't save spoiled milk; that sour milk scent will not be masked.

Natural breeding does not take long and is unceremonious. Once a doe is bred she rejects her suitor's continuing advances. There are ultrasound devices available if you want to know for certain if a doe is bred and how many kids she will have. Often you can find someone to come to your farm and "preg check" your does for a fee. Some labs run tests on the animal's blood to check for the presence of pregnancy-specific proteins produced by

A Nubian buck. This registered animal is a fine example of a mature male. His stature, body conformation, and overall condition are prime for breeding the herd of does. A buck can breed many does in a 24-hour time span.

Spots and speckles make for a showy goat. Some goatkeepers breed specifically for spots, working through genetic selection and rebreeding those with the most color.

the fetus. This information can be helpful, especially if you only have two or three does and are dependent on the milk. If the doe is not bred, she will continue to cycle for several months. It is rare that this would happen unless she is too old to conceive or the buck is sterile.

Artificial insemination is also an option to consider. While this practice is not as common to use with goats as it is with cattle, a trained professional can facilitate the process with good results. Semen is available for purchase, and this is a way to bring superior bloodlines into your herd. The beauty of this type of breeding is that a buck does not have

to be maintained for 365 days a year. Also, the date of conception is a known fact, which is helpful when charting a due date.

The doe shows signs of pregnancy, including the enlarging of the udder, toward the end of gestation period. Her belly becomes noticeably larger, and her hair coat starts to fringe out, almost like a little skirt, flaring out over her ample belly. She becomes more labored in her walking and getting up. During the last few days, before her kids are born, it is quite common to hear her talking to them. She makes small noises, as if she is giving preliminary instructions to her unborn.

For most does, the birthing process is uneventful. Dairy goats are generally good and devoted mothers. Check the calendar for the suspected breeding date and watch for

Superior Males

When I was first learning about goats, many experienced goatkeepers informed me that the buck sets the quality of the herd. In other words, a grade doe can produce superior offspring if she is bred to a superior male. I was told that was the place to spend my money—on the buck. I did this and purchased a magnificent registered Nubian male. He appeared to be aware of his pedigreed linage, and he was quite a proud being with a regal posture. I had grade does (nonregistered, mostly mixed breeds) that produced beautiful healthy kids. Some goatkeepers may prefer to invest in registered animals. Often these animals can be traced back to royal lineage.

Line breeding refers to the breeding of daughters back to fathers. This can serve to strengthen desirable traits. However, it can also have a negative effect by enhancing genetic flaws. There are breeders on both sides of this fence. Research their opinions before making a decision.

Closed herds are becoming more prevalent in an attempt to reduce disease brought onto the farm by new animals. In a closed-herd system, after the initial breeding stock is selected, replacements are bred from within.

This buck probably has mischief on his mind as he stands guard over the entrance to his shelter. Bucks become quite protective and territorial during mating season.

the 145 to 155 day mark. The doe may begin a slight discharge from her hindquarters; some do and some don't. When the time comes, she may pace around the field with two sets of little hooves protruding from the birth canal. This is typical, so don't be alarmed. If all is going well, soon a small nose and head will become visible, and it won't be long until she will push the new babe into the world. She will clean it, remove the mucus, and eat the sack that formed the protective bubble around the kid. Don't be surprised if in a few minutes, another kid begins his way into the world. The doe may lie down.

If I know a goat is going to kid, I typically bring her into a kidding area and confine her, especially if the weather is cold or rainy. A chilled kid does not do well. However, some goats that will resist this effort and refuse to give birth in any kind of stall. Remember their nomadic spirits. Some wait until they are out in the field and simply come back a few hours later with their babies wobbling behind them.

Problems that can occur are spontaneous abortion, breech births, toxemia, ketosis (a metabolic imbalance that can be associated with pregnancy and the postnatal period), and infection. A farmer friend keeps a video camera in the barn so she can turn on her goat cam and observe her animals at any time of the day or night. A baby monitor picks up the sounds of a nanny when she goes into labor—if the nanny is one who changes the usual pitch of her voice or nervously bleats during kidding. Barn checks become frequent, especially on those cold nights when you wake up and wonder about what is happening in the barn. Keep a flashlight and a barn coat handy, for these kinds of events are a part of being a goatkeeper.

I never thought I would serve as a midwife, but there are times when a goat needs assistance, and if you happen to be the one there, you are elected. Don't be overenthusiastic when deciding whether to pull a kid. Nature will often take its course—if you wait it out—and this is the best circumstance for all involved. Should it become necessary to pull the kid (if the mother exhibits signs of great distress, for example), the person with the smallest hands will have the best luck and cause the least amount of discomfort to the animal. Wash your hands thoroughly, wear rubber gloves, and then use a lubricant. Insert your hands into the birth canal gingerly, and try to feel how the kid is positioned. I usually close my eyes and feel my way around, getting a mental picture of the kid, its legs, and torso. If more than one kid is present, they may be entangled, so you will literally be sorting out legs and bodies. Try to stay calm, talk to the doe, and work the babies out, one at a time. Check the umbilical cords and if they are looped around the neck of a kid, remove them immediately. Once the kids are on the ground, wipe the mucous from their noses and mouths and give them to their mother. She will take care of them from there, and the less you do, the less risk there is of rejection. Keep an eye on mamma. She may be exhausted, but it is important for her to get up fairly soon after birth. Give her plenty of water, which she should drink enthusiastically.

Check the kids, and dip their umbilicals in iodine to prevent naval ill. Within a very short period of time, they will be up and running, hopping, and skipping. Such wonderful little creatures. Hold one and look at its tiny little hooves, its bright eyes, and zest for life. Kids are truly the cure for midwinter

blues. The doe must deliver the afterbirth within a short time after kidding. If she does not, she risks facing a toxic condition, infection will set in, and the doe will die. If the animal is confined and you can keep an eye on her, then you can be sure this event has taken place. It is much more difficult to know if she has delivered the afterbirth if the animal is out on pasture. The doe will most likely eat the afterbirth, so unless you are there, it is hard to know for sure. I assume that all has gone well unless I start to see the goat looking downcast, observe her kicking at her belly, or find her refusing to get up. At this point, she will start to run a fever (normal temperature is 101.5° to 103.5° F), and things usually go downhill from there. I call the vet or risk losing the animal. This event is rare, so don't go into the delivery expecting the worst. Simply be prepared and know the signs if things do not go as planned. If need be, the vet will administer antibiotics and a medication that will cause the uterus to convulse and expel the remaining fluids. Steroids are often injected to jump start the immune system and encourage healing. Watch and learn.

Milk fever (hypocalcaemia) is a condition that can occur in pregnant does, newly freshened does, and even dry does. The doe may present with symptoms such as shivering, poor balance, lethargy, and poor muscle control. Often a decreased appetite, rumen dysfunction, and a drastic drop in milk production may also be present. The term "fever" is misleading, as the cause of this illness is a calcium deficiency. Calcium supplements are available to counteract this deficiency. Consult a veterinarian for dosage and recommendations.

If a mother rejects a kid, then you will have a project. Sometimes it is possible to graft a kid onto a more willing mother. The choices are to let nature take its course and the kid will die or for the goatkeeper to become a surrogate mother for a while. The kid will need colostrum, the antibody-strong first milk received from the mother. Either milk it out from the mother, take some from another goat, or use a cow's. Some commercial replacements are available, but the real thing is always best. If an animal does not receive colostrum it will fail to thrive and likely have a short, unhealthy life.

Generally kids should have about 1 ounce of colostrum per pound of body weight. Many experts recommend that colostrum and milk for babies should be pasteurized to prevent the transference of disease. Screw a small nipple, available through the goat supply houses, on to a plastic soda bottle. Observe the kids who are with their mothers; they have tiny tummies and eat small amounts quite frequently at first. Feed the kid the colostrum first, limiting the amount of milk to 1 or 2 ounces for the first few days. Do not allow them to overeat or they will have scours. As they grow, increase the amount of milk and decrease the number of feeding times per day. Allow 12 ounces or so before bottle feeding ends after about 3 months.

If you have a crowd to take care of, check into a milk bucket with multiple nipples. Simply fill the pail, and place the kid's mouth on the nipple. Work the nipple a little so milk will come out, and the kid will quickly get the hang of things. Make sure if you have a weak kid to ensure that he or she gets a fair share. If a kid is very weak, he or she will need individual attention.

Kids will need milk for about 3 months. If they are left with their mother, she will tend to weaning them. If feeding is your job, then

This photo shows the LaMancha "elf ears," the trademark of the breed. These goats tend to be light in color, gentle, and less talkative than some of the other breeds.

at about 10 weeks, start to cut back on the amount of milk presented at each feeding. By week 11, go to one bottle per day. During this time the kids will be eating grass, foraging, and eating hay (if provided) and small amounts of a goat ration. A number of manufactures make a sweet feed that is properly balanced to meet the animals' nutritional needs. Salt and mineral blocks are recommended and should be placed where the animals will locate them as needed.

Care of Kids and Does

This is a good time to note that veterinary medicines are available at local farm stores. Certain vaccinations are recommended when the kids are born. Get these recommendations from your veterinarian, and learn to administer the injections. Typical protocols include the vaccinations for clostridial diseases, specifically enterotoxaemia (type C and D) and tetanus. Other vaccines to be considered are those for sore mouth, caseous lymphadenitis (CLA), abortions (e.g., vibrio, chlamidia), and rabies. If something unusual happens or if an animal does not respond to treatments within 8 to 12 hours, call in a professional. It is false economics to save money on a vet bill only to

A lady in waiting. In full bloom of pregnancy, this doe is out browsing in the spring growth. Noting her girth, twins are likely. Twins are a common occurrence with the dairy breeds.

lose an animal. Always remember to dispose of used needles properly and do not reuse the needles or you risk transferring disease from goat to goat.

If a kid is rejected by its mother, chances are it will be found after it has chilled. Take the animal to a warm room and wrap it in a towel with a hot water bottle. Massage it to bring the body temperature up. Place a small amount of colostrum in a heavy pot (or double boiler) and warm it slightly. Do not warm the colostrum in a microwave. This will destroy the active antibodies. Pour the colostrum into a bottle with a nipple and work with the kid

until it sucks. Move the bottle back and forth in its mouth if necessary. Sometimes it is very difficult for a weak kid to suck from a bottle. In this case, draw up the colostrum into a 20-millileter syringe (with no needle). Push the plunger down slowly, allowing very small amounts of colostrum to go into the kid's mouth. Massage the throat so it will swallow. Aspiration of milk can be a risk. Make sure to go slowly while introducing the milk.

Ask your veterinarian to show you how to intubate a kid. This involves a feeding tube and a syringe. Purchase these ahead of your need for them and make sure you are familiar with this technique before the need for it arises. Chances are, if the kid is weak and chilled there will be a short window of opportunity. This may sound like a complicated technique to learn. It isn't, and once you are comfortable with the procedure it can save many a weak kid, so it is well worth the effort. Occasionally, if a kid is nonresponsive to general warming techniques, there is one last-ditch effort that can revive a nearly dead animal. Fill a sink with warm (not hot) water and place the kid in it. Be careful to keep the head above water. Gently massage the kid and try to stimulate it. As it begins to show signs of responsiveness, dry it off and then try to get it to take warm colostrum. If all goes well, it should take off from there.

Milking

If you have not milked before, be prepared to feel quite awkward. If you can, milk someone else's goat that has been trained and knows what to do. Most of the time the trained animal climb up on a stanchion (an elevated milking stand with a feed box). Grain is placed in the box so the goat can eat while she is milked. In the beginning, don't feel badly if it takes a while to milk the animal out all the way. You will gain speed and confidence as you go along. Milking is a twice a day task, best performed at 12 hour intervals. This is a twice-daily event come rain or shine—truly for better or worse!

There are two basic ways to milk: by hand or by machine. Each method has its benefits and consequences. Hand milking is simple, basic, and creates an intimate bond between man and animal. For some purists, this is the only way to go. If there are more than a few animals to be milked, hand milking will require some additional hands, as it is labor intensive. Milking by machine offers convenience and ease, especially if your herd is larger or your help is limited.

Nubians, well known for the length of their ears, are more talkative than other goats. Personable animals, they also give the milk with the highest butterfat content.

Milking by Hand

For some, this instruction may be elementary. However, if you have never milked an animal before, it can be quite intimidating! When I first sat down by the goat, I felt extremely awkward and proceeded to wear both myself and the goat out in the process. I was unsure and nervous, so the animals immediately sensed that and behaved in a similar manner.

Goat bells are common, and there is nothing like the sound of a herd going through a field with their collar bells tinkling. Mountain goatkeepers once used bells as a way of keeping track of their animals. This Nubian doe has a good body conformation and a well-attached udder. These are two of the things to look for when selecting a goat for your farm. A slick coat is also an indicator of good health.

For those who have milked all their lives, it is probably laughable that someone would struggle with this most basic of country skills, but believe me I did. I finally asked for the advice of a neighbor, a long time dairy woman. She explained the milking process to me as follows: Think of the mammary gland as a cistern. When you draw down on the teat the cistern releases the liquid. In between the drawing of the milk, the cistern fills back up and the milk is there when drawn down again. This helped me to have a visual image of a bag filling up with fluid, then releasing as I drew milk from the teat. Watch how a baby goat goes about getting milk from mamma. First he nudges the udder, aggressively, to let mamma know it's time to get down to business. Then he begins to suck. Mamma lets the milk down and dinner is served.

Before bringing home your first milking goat, buy or build a stanchion. A stanchion is an elevated platform with a head stall and a feed box. The height of the stanchion takes some of the stress off the farmer's back. The goat will stay busy munching feed from the feed box, and the head stall has a bar that will drop down over the back of the goat's neck with gentle pressure to keep her in place while milking. Place the stanchion somewhere that will be convenient to the goat yard. A small area with a concrete floor is ideal for sanitation purposes. Potable water is a necessity. Always hose down the area and sanitize before and after milking. It is a luxury to have an entrance door on one end of the building and an exit door on the other, leading back to the pasture. If you are milking more than one animal, bring them into a small pen outside

Off-Tasting Milk

Off flavors in milk come from several sources: if a buck is allowed in with the does, they will pick up his musky scent; a goat that is ill can present with off-flavored milk; and an animal that consumes wild garlic or onions will pass that taste to the milk. It is quite unpleasant. Quick cooling of the milk will preserve the quality and inhibit bacteria growth and off tasting flavors. Goats need fresh drinking water to keep milk flowing and tasting sweet.

Milk may also have an off taste if the goat is ill. Mastitis is not always detectable to the eye. Test are available to check white blood cell counts in the milk. These cells are always present in the milk, however when the count goes beyond an acceptable level mastitis is diagnosed regardless of the visual composition of the milk. The affected goat will still require milking each day but the milk should be discarded. Mastitis is treated with antibiotics, often inserted as a paste through the opening in the teat. For commercial herds, there are companies that will run a milk check on a regular basis. These companies attach instruments to the milking equipment that measure overall output, butterfat content, overall solid content, and cell counts. This is good and necessary data if the goal is to go into commercial cheese production.

It goes without saying that any milk that shows signs of mastitis should be discarded. Any milk that smells sour or acrid should be discarded as well. These are warnings that the goat is ill and medical intervention is required.

the milking parlor where they will await their turn on the stanchion. Routine is important and establishing this in an animal's mind is important.

Of course, as humans, we introduce an artificial environment: the milking stanchion, our touch, and various things that can put an animal on edge. There are times, especially with a first freshener, when the animal will simply refuse to let her milk down. This reflex action is within her control, but a few things can be done to dissuade her from withholding her milk. First, talk to her and try to put her at ease. Gently bump the udder, mimicking the motion of a kid. Allow her to eat while in the stanchion and accept the fact that the first few attempts may not be entirely successful. Nature will take its course and as that udder fills, eventually the goat will come to an understanding that relief comes with milking. Some goats break easily to the stanchion and others are a challenge. Don't think I am totally crazy, but sometimes I find that singing to the doe actually helps her let her milk down. Maybe they associate this racket with mistreatment and simply want to hasten the entire process!

You'll need to purchase a few items to get set up for milking by hand. A strip cup— to catch the first steams of milk—will have a screen to quickly show any signs of clumping in the milk. Also recommended are a hooded stainless steel milk pail, teat dip, dairy sanitizer, paper towels, disposable gloves, and containers to store the milk.

Always remember basic sanitation when you are milking. Keep the hair trimmed around the udder. This is the perfect place for debris to gather and then fall into the milk pail. Wash your own hands before milking. Disposable gloves are recommended. As the goat enters the milk parlor or climbs onto the stanchion, use a strip cup to capture the first milk stream or two and check for signs of mastitis (stringy, off smelling, or curdy-looking milk). These first streams of milk are bacteria laden. The teat is an orifice that will pick up dirt and bacteria between milkings. The first flow of milk has a high bacteria count, so discard that and your overall bacteria count will be lower. If you consider that the milk rests in that teat all day, warm and exposed to all sorts of elements, you will understand why discarding this initial offering is important. Then dip the teat in sanitary teat-dip solution. Completely cover the teat with the dip. A common error is to dip only the side facing the milker, so make sure the dip encompasses the entire teat. Dip cups are available for this purpose. The last step is to thoroughly wipe the teats, ridding them of the sanitary cleaner. A clean rag should be used for each animal, otherwise, there is the risk of pathogen transmittal from one animal to the other. These rags should be disposable, or great care should be taken in their washing, as the pathogens can survive mild washing temperatures. From there the milker may proceed with either hand or mechanical milking.

To milk, grasp one teat in each hand and compress your fingers in toward your palm. It takes a little while to develop the muscles in your hands, particularly those in between your thumb and first finger. Get a rhythm going milking one teat, then the other. Swish, swish, swish, swish. . . . The stream will hit the bucket after a few times of milking, and you will become an old pro. Sometimes the goat will have a very small opening in its teat and will remain difficult to milk for its entire life. There is no cure for this, and if you are hand milking, it is a royal pain.

Care of Milking Equipment

The production of high quality milk requires strict cleaning and sanitizing procedures for all equipment that comes in contact with milk. Cleaning and sanitization can be done manually or mechanically depending on the type of equipment used on the farm.

The basic steps in cleaning and sanitization are as follows:

• Immediately after milking, or removal of milk from equipment, rinse the equipment with lukewarm water before the milk dries on the surface.

• Prepare a detergent solution according to manufacturer's specifications, making sure the water temperature meets or exceeds the minimum recommended temperature. If manual cleaning is employed, brush all milk contact surfaces thoroughly. All milk contact surfaces that are not cleaned by mechanical cleaning or circulation cleaning must be brushed.

• Rinse detergent from tank with tap water. Preferably, an acidified rinse should be used to prevent the accumulation of milkstone.

• Drain rinse water from all equipment.

• Immediately before using the equipment sanitize with an approved dairy sanitizer. According to the *Dairy Goat Production Guide* from the University of Florida's Animal Sciences Department, either chlorine or iodine can be used at concentrations of 200 and 25 ppm, respectively.

• Wash down the milking parlor, stanchion, floors, and walls with a sanitizing solution. Bacteria are there waiting for an opportunity to multiply. A good sanitation routine can keep them under control.

Milking by Machine

If you have more than a few goats, a milking machine is worth consideration. Electric machines are available from a number of suppliers. Some units milk one goat at a time; pipeline units milk 25 or more at a time. The small units plug into electricity; larger ones work off a separate compressor. In most units, a pulsator will cause the teat cups to inflate and deflate, which puts pressure on the teat and creates a vacuum, sucking the milk from the udder. Milk can be directed into cans for individual animals or into bulk tanks for large herd pipeline milking systems. Large milking facilities are often constructed around a concrete pit, and the workers milk from the pit below the animals. The goats enter the parlor, usually 12 at a time, and eat their rations while the milking machines are put in place. A great number of animals can be milked in a relatively short period of time when using this type of system.

Expect the animals to be leery of machinery in the beginning. They will get used to it, but the training period can be pretty exciting for the first few milkings! Remember, the animals will not understand the reasoning behind this new machine, and soothing words from you will help them become less nervous about the whole situation.

Milk the goats until they are thoroughly milked out, but never wring the teat to get the last drops. Wringing can cause damage to the blood vessels. After milking, spray the teats again with iodine-based spray, and turn the animal back out to pasture. Thoroughly clean all milking equipment with an approved dairy sanitizer.

Quick cooling of milk is essential to keep bacteria counts low and to ensure good-tasting milk. If you have milked by hand or with a small milking system, strain the milk through a strainer with a disposable pad. Then chill the milk as soon as possible. Placing the milk container in an ice water bath is the quickest way to bring the temperature down. The desired chilling temperature is 35°C. Cap the container only after the milk is thoroughly chilled. It goes without saying that the jars used for storing milk should be thoroughly sanitized.

How to Dry Up a Doe

You are probably thinking why would I want to dry up a doe? If the plan is to rebreed the doe, then she will need time to rebuild her strength and stamina. Milk production takes energy. Kid production takes energy, too.

Most goats will start to slack off in their milk supply after about 10 months. There are exceptions to this rule, but we will go with the usual chain of events. If the doe is to kid annually, then we would breed her in July or August, when she first cycles into heat for an anticipated kidding date of February or March. The gestation cycle is about 5 months. We continue to milk the doe for the first 3 months of her pregnancy. The last 2 months she should be dry so her energy goes to the unborn kid, not to the milk supply.

If the plan is to have a fresh goat in March, then toward the end of October start to dry her up. The first thing to do is cut back on her grain. Grain fuels milk production. Reducing her grain is the most difficult part of this process. She will be expecting grain when she jumps onto her stanchion. Just give her a very small amount to keep her in her routine.

Along with the reduction in grain, milk her out a little less each day. The milk supply is based on supply and demand. If you milk out all the milk, the doe will simply produce enough milk to refill the demand. So by leaving some milk in the udder, the milk supply will start to naturally lessen. Her udder will become taut, and she won't be too happy during this process. Milk her out to the point she seems comfortable, then stop. Continue to do this for about two weeks. You will notice a decrease in the amount of milk each day. After following this routine for the two-week period, then milk the doe out only once a day for another week. The next week, milk her out every other day if needed. By this time, nature should have taken its course and her udder will be nearly empty of milk and she will no longer produce milk until her next kids are born.

Milking Through

As a farmer, you may look forward to staying out of the milk barn for a couple of months in the winter, but when your doe is out of milk, if that is your primary milk supply, there will be an empty spot in the refrigerator. Occasionally a doe is superior in her milk output. While this is the exception rather than the rule, some goats can remain in milk for more than the usual 10 months or so. Some goats may remain in milk for 2 or even 3 years without refreshening.

Milking through is advantageous for several reasons. The first, of course, is the continuous supply of milk. Second, the doe doesn't have to go through another pregnancy or kidding. With any pregnancy, there are risks. When a doe doesn't experience the stress of pregnancy, she will most likely live longer, too.

How do you know if your goat has the potential to milk through? Look for higher than usual milk production. Strong milkers will taper off but not to the degree of the other goats with lesser milk supplies. The breed with the heaviest milk supply is a Saanen, and they are the most likely candidates for milking through.

❦ Maintaining the Herd ❦

Record keeping will also make your life easier in the long run and is a tremendous help in keeping track of the needs of the herd. Computer programs will generate reports, send reminders about vaccination dates, and note other points in your maintenance program. Animals are tracked by number (or name), then initial data is entered (age, purchase date and source, breed, overall health, date of vaccinations, date of worming, etc.). If you don't care for the computer tracking version, create your own system with pen and paper, recording your dates and creating a chart to remind you of services provided and schedules for maintenance. Even if you only have a few goats, tracking will be very helpful. If you breed goats for sale, buyers also will appreciate the information pertaining to pounds of milk per day, kidding rate, age, immunizations, and other maintenance records. Whether you name your goats or give them a number, make sure each has a tag. This will help in identification and if you have to leave directions for someone.

Disbudding of kids can be somewhat traumatic for the kid and the goatkeeper, but completing this operation will prevent future problems. Again, have a professional disbud your kids in the beginning. You can learn the basics and purchase a disbudding iron after you learn the process. Always make sure to disbud goats during cool weather. Flies are always looking for a host and will attempt to lay eggs in an open cavity.

Fiber and meat goats are not usually disbudded. However, after witnessing an animal with a severe tear in her abdomen after a goat argument, I became a believer in the disbudding dairy goats. The horn buds are cauterized (literally burned) with either caustic paste (not recommended) or a disbudding iron. Some people swear by the paste, but I have always found issues with this method. Sometimes the horns continue to grow but are weak; a knock on the head will disengage the horn from the scull, then blood pumps freely. With a disbudding iron, once the procedure is done, it is done and there are no more concerns. Using an iron does cause the animal pain. However, the discomfort is relatively short lived, and in the long run, dairy goats without horns are easier to manage and do not injure each other with their horns.

Castration of males is mainly done with dairy goats and fiber animals where too many males can quickly become a problem. For my part, I use a banding system that works well and is not too traumatic. This procedure is done 7 days after birth. A tool is used to spread the band large enough to slip over the testicles and then the band is placed. The blood supply is shut off, so the testicles eventually drop off. When this procedure is done at the proper time, the kid will be largely unaffected and not in a great deal of pain. An older kid will need to be taken in for a surgery. As with disbudding, castration should take place during the cold months a few days after birth.

Castration creates a wether, which can run with the females, put on weight, and either be used for meat or for a brush control forager. A male dairy goat will gain weight and the meat is fine for eating, but this animal will not have the same bulk as another breed grown specifically for meat.

Urinary calculi is a condition that occurs mainly in wethers, though it may also been seen in unaltered males. Similar to kidney stones in humans (phosphate salts become concentrated and form crystals which are difficult to pass through the urinary tract), it can be exacerbated by castration because it affects the hormonal balance and thus the growth of the genitals. Other factors include feeding concentrates that are excessive in phosphorus and magnesium. Lack of water intake by the animal can also add to the likelihood of problems. Check with your feed manufacturer for the proper formulation to avoid excess mineral contents.

Mastitis is not always detectable to the eye. However, a goat may begin to exhibit symptoms before physical signs are present in the milk. Often, the goat will begin with a downcast spirit and become less active than normal. She may run a fever. Purchase a thermometer from an animal supply house. Normal temperature ranges from 101.5° to 104° F. The udder may become red or hot and the goat may kick at the udder. Tests for somatic cell counts (an elevated level can indicate mastitis) can be used to check for white

blood cell counts in the milk. When the count goes beyond an acceptable level, mastitis is diagnosed regardless of the visual composition of the milk. (Note: The somatic cell count is an indicator of milk quality. A high cell count is an indicator of pathogenic bacteria.) The affected goat will still require milking each day but the milk should be discarded. The condition is treated with antibiotics, often inserted as a paste through the opening in the teat. General antibiotic treatment may also be recommended. Check with your vet for protocols.

As with human maladies, many conditions can complicate goatkeeping. The

In addition to forage, pasture grasses provide camouflage.

Merck Veterinary Manual is an excellent resource that includes diagnostic information, treatments, and prescriptive recommendations for all medical conditions affecting goats and their offspring. I have dealt with both small herds and a herd of over 300 animals, and through personal experience, I have found that the incidence of disease is relative to the herd size. In a larger population, I found an increase in the occurrence and frequency of disease, as well as more unusual diagnoses, such as tetanus and other conditions I had not seen in a small herd. Quarantine is a practical means of preventing the introduction of new health concerns when bringing animals into a herd. Some goatkeepers keep a closed herd, producing their own replacements to prevent disease and other health issues from occurring.

Cheesemaking with Goat Milk

*N*ow what to do with the milk? Goat milk is wonderful and often preferred over cow milk. It makes the best ice cream, macaroni and cheese, and scalloped potatoes. I love it in everything from pancake mix to salad dressings. Of course, cheesemaking is the ultimate goat-milk creation. For this book, I focus on fresh cheeses. They are easy to make, offer a quick rate of gratification, and are the best way to learn this new skill.

Cheesemaking is a practice known all over the world. It is largely influenced by the region, type of milk available, climate, and historical knowledge. To be sure, wherever you find a dairy animal you will also find some type of cheese. Cheese can be made from cow, goat, sheep, mare, yak, water buffalo, camel, and other lactating species. Each type of milk is known for its various attributes and contributions to cheesemaking. Goat milk in particular is known for that specific tang that comes from caprylic acid, a component only found in this type of milk. That acidic bite assists the cheesemaker in creating a product that naturally has a flavor component all its own.

✤ Basic Cheesemaking ✤

Making fresh (unripened) cheese is very easy. First, the milk should be pasteurized (see sidebar, page 48). Then beneficial bacteria, known as a culture, is be added to enhance the natural flora. For the recipes included in this book, the culture is Mesophilic (MA), which works at low temperatures. Calcium chloride, an additive, may also be required to firm up the goat curd, which can become weakened in late lactation or in the heat of the summer. Rennet, an enzyme that causes the milk to form a gel, will also be added. Rennet can be derived from an animal, vegetable, or chemical source and is similar to the consistency of yogurt or custard. Salt also plays a strong role in cheesemaking, adding flavor, assisting in the dehydration process, and adding a preservation component. The basics of cheesemaking are brought down to two simple elements: curds and whey. The curd is the semisolid substance; the whey is the liquid that is shed from the curd.

Brie, also known as the Queen of Cheese. The addition of Penicillium candidum *brings forth a bloomy rind with a hint of mushroom. Serve at room temperature.* Jiri Hera/Shutterstock.com

Milk is a complex substance containing fat, proteins, lactose, casein, minerals, and other inert components (cells), but it is also about 90% water. As cheesemakers, we want to get rid of a large percentage of that water while retaining the fat and solids found in the milk. Expect the yield of a soft cheese to be about 12 to 15% (or 1.2 to 1.5 pounds) cheese per gallon of goat milk. A hard cheese will have an expected yield of approximately 1 pound per gallon of milk. You will note that this rate of loss is about 90%, which correlates with our original figure of the milk's composition of 90% water. Some moisture is retained—more so in fresh cheese—which makes a creamy spread. Due to the high moisture content,

fresh cheese has a short shelf life. It is truly best when made and eaten within 2 days, but it can be kept for 10 to 14 days.

When working with milk, the industry standard is to work in pounds rather than gallons. One gallon of goat milk weighs approximately 8.5 pounds. Therefore when a goatkeeper is asked how much milk a particular goat produces per day, the answer would be in pounds rather than gallons.

About Chevre

One of the things most enjoyable about chevre is that it will complement almost any food or drink. It is very light, so any herbs or flavors

continued on page 48

nav
nav
nav
nav
nav
nav
nav

continued on page 48

Cheesemakers' Shopping List

Cheesemaking supplies are available from suppliers across the United States. (See the resource section, page 187.)

• Fresh milk. Recipes will specify whole (with the cream) or skimmed (with part of the cream skimmed from the milk). Make note of pasteurization recommendations for all cheese that is to be aged less than 60 days.

• Culture. All recipes in this book use an MA (Mesophilic) culture, which is very versatile and is used in cheese cooked at low temperatures. MA culture thrives at temperatures ranging from 80 to 86° F. I prefer a direct vat inoculant (DVI). A DVI is more or less an instant culture, which can be added to the milk and used shortly after innoculation. Other types (mother cultures) require an incubation period in a controlled environment.

• Rennet. This liquid or tablet is an animal, vegetable, or chemical base. You can make a choice based on preference, however, the tablets keep longer, up to 1 year in the freezer, and these are vegetarian based. One tablet is the equivalent of 1 teaspoon. Fractions can be figured accordingly: $\frac{1}{8}$ tablet = $\frac{1}{8}$ teaspoon, $\frac{1}{4}$ tablet = $\frac{1}{4}$ teaspoon, and so on.

• Calcium chloride. This compound is sometimes required to get a firm set. Goat milk is fragile, and it will change through the lactation cycle. If the milk refuses to set, add calcium chloride in the same quantity as rennet to create a firmer set and therefore a firmer curd.

• Citric acid. This a weak acid is made from crystallized fruit sugars and is often used as a preservative. It also is used in mozzarella where it serves to acidify the milk and takes the place of a culture.

• Lipase. Use this animal-based enzyme to add flavor to Italian cheese. There are three varieties/strengths: calf, which is mild/picante; kid, pecorino; and kid-picante, sharp.

AlenKadr/Shutterstock.com

• Salt. Look for noniodized, sea salt, or kosher.

• Dairy thermometer. Make sure to purchase a thermometer that registers 0 to 220° F.

• Cheese molds. Purchase the recommended sizes through cheesemaking supply companies.

• Cheesecloth. Try the disposable types available from cheesemaking supply companies. (Tip: Even though they are marketed as disposable, these cheesecloths are actually reusable. Simply wash, rinse, and sanitize, then hang to dry.)

• Nonchlorinated water. This element is essential for diluted rennet (chlorine can affect the effectiveness of the rennet).

• Other supplies: Flour-sack dish towel or cheese draining bag, heavy cooking pot large enough to hold 1 gallon of milk, double boiler, colander, measuring spoons, slotted spoon, ladle, knife or spatula for cutting curds, draining board to place over sink, timer, and a heat source for warming milk.

Chevre: The Classic French Cheese

As you make these fresh cheeses, you will find literally dozens of ways to use them. I dread the day when the goats go dry and there is no chevre until spring! This cheese becomes a staple, and nothing else quite fills the bill. There simply is no substitute for fresh chevre.

Ingredients

1 gallon of fresh goat milk (pasteurized)
2 drops of liquid rennet dissolved in ¼ cup nonchlorinated water
⅛ teaspoon Mesophilic DVI MA culture
½–1 teaspoon noniodized salt to taste
Yields 1¼ pounds

Equipment

Heavy cooking pot (large enough to hold 1 gallon of milk)
Dairy thermometer
Slotted spoon
Ladle, string
Colander
Flour sack, dish towel, or cheese draining bag

1. Pour the goat milk into the cooking pot. Heat the milk slowly to 86° F.

2. At 86° F remove the pot from heat.

3. Sprinkle the culture over the top of the milk and gently stir, making sure the culture is dissolved and well integrated into the milk. Allow this mixture to sit for about 45 minutes so the culture has time to develop.

4. Add the rennet mixed in water to the pot and stir, coming up from the bottom of the pot, until the culture and rennet are mixed thoroughly into the milk. (Stir gently for about 1 minute.) Let the mixture rest, covered with a cloth, in a warm place (70° F) for 12 to 18 hours. The gel will thicken to the consistency of yogurt while it is resting.

5. Line a colander with a flour sack or tea towel. Place the colander in the sink. When the gel has thickened, gently transfer the gel mass, now called the curd, into the lined colander. Keep ladling until all the curd is in the colander. The leftover liquid is called whey, which is a waste product—although chickens and pigs love it!

Chevre, possibly the most elegant of all cheese, is simple to make and a complement to lighter fare, such as salads and fruit. HLPhoto/Shutterstock.com

6. When the curd is all in the colander, gather the tea towel corners and tie them with a string to form a bag. Hang the bag over the sink so the whey will drain freely.

Two things are happening while the curd drains: acid is developing, so the flavor of the cheese is coming to life, and the moisture ratio of liquid to solid is dropping. Both of these actions change the consistency and the stability of the finished product. Chevre is meant to be soft, so the moisture level will remain high. Because the high moisture content makes chevre less stable than other aged or hard cheese, it should be consumed within a few days after it is made. (This cheese will not improve with age; it is meant to be eaten as a freshly made product).

7. Allow the curd to drain for 12 hours. Then remove it from the bag and place it in a bowl. Work in the salt. Salting has a number of purposes in the cheesemaking process. It adds flavor, promotes the shedding of moisture, and retards bacterial growth. Cover and refrigerate.

Chevre hangs in a bag to drain. A no-nap dish cloth is the perfect weight for draining. The whey will slowly drip from the solid, leaving a velvety soft curd. This draining technique has been used for centuries.

continued from page 43

Pasteurization

Please be aware that federal regulations require the pasteurization of raw milk for making cheese that is to be aged for less than 60 days. This chevre recipe is for a fresh cheese, so the milk should be pasteurized as follows. Always remember children, pregnant women, the elderly, and those with compromised immune systems are warned against using raw milk.

Equipment
Dairy thermometer
Double boiler or 2 nesting pots

1. Pour the milk into the smaller of the two pots. Run 3 inches of water in the bottom of the large pot, and place the small pot inside the larger one.

2. Slowly heat the milk to 145° F, and hold the temperature there for 30 minutes. Stir the milk gently throughout the process to make sure it is evenly heated.

3. Remove the milk from the heat source, and place the pot in a sink filled with ice to bring the temperature down as quickly as possible. If you are ready to make this chevre recipe, bring it down to 86° F and begin your cheesemaking. If the milk is to be stored after pasteurization, bring it down to 40° F. Refrigerate until use.

added to it will not become overpowered. Fresh chives, garlic, thyme, lavender, and the French blends are perfect accents for this cheese. Chevre adds a touch of elegance, and everyone will assume you are an expert at cheesemaking once they taste this most simple of cheeses. I literally use this cheese at breakfast, lunch, and dinner—and often for dessert.

Chevre does not melt, though after it sits for a few days it can be crumbled and served on pizza or warm French bread. One of my favorite ways to serve chevre is with fresh snow peas, straight from the garden. Simply slit the peas at their seam and gently tuck in bits of the cheese. Sprinkle with freshly ground black pepper and a bit of salt. Create a rustic salad complement by adding chevre to an heirloom cherry–type tomato. Cut an "x" in the center of the tomato, add chevre, and top your creation with a small basil leaf.

For a very simple dessert, make the chevre recipe, omit the salt, and add ½ cup (more or less to taste) of prepared lemon curd. Stir the lemon curd into the chevre after draining the cheese and before refrigerating it. This makes a light dessert that is perfect served with a chewy gingersnap. Garnish it with a bit of lemon zest or lemon balm.

For a cheese that has a bit firmer consistency, try this recipe. The classic format for this cheese is a crottin mold, which makes small round discs of cheese. The suggested size for this mold is 2½ inches wide by 4¾ inches tall with a solid bottom. The molds are made of food grade polypropylene and are worth the small investment, as they will last for years. Fill the molds to the top with curds and whey. After 12 hours, the whey will drain away and a small cheese, about 1½ to 2 inches tall will be left in the mold.

Fresh Goat Cheese

Ingredients

1 gallon whole goat milk (pasteurized)
⅛ teaspoon calcium chloride diluted in ¼ cup cool, nonchlorinated water
⅛ teaspoon Mesophilic DVI MA culture
⅛ teaspoon liquid rennet diluted in ¼ cup cool, nonchlorinated water
Salt
Herbs, optional
Yields 1 pound

Equipment

Heavy cooking pot (large enough to hold 1 gallon of milk)
Dairy thermometer
Slotted spoon
Ladle
4 crottin cheese molds
Draining rack

1. Pour the milk into the heavy pot, and add the diluted calcium chloride.

2. Warm the milk to 86° F, and remove it from the heat.

3. Sprinkle the culture over the top of the milk; gently stir, making sure the culture is dissolved and well integrated into the milk.

4. Add the rennet mixed in water to the pot and stir, coming up from the bottom of the pot, until the culture is well integrated into the milk. Let the mixture rest, covered with a cloth.

5. Test the milk for signs of coagulation in about 20 minutes. To do this, simply slice through the gel (curd) with a long knife and look for a distinct separation. This is a clean break. If the curd is not well defined at this point and there is not a distinct slit in the gel when it is tested, wait 5 more minutes and slice through the mass again. If it looks like yogurt, it is not ready. Test again 5 minutes later. The break should be achieved at 25 to 35 minutes out.

   Achieving the clean break and cutting at the proper time is one of *the* critical points in cheesemaking. Cut too soon and the curd will not be well defined and the result will be very liquid. Cut too late? The result will be tough curds, ones that do not want to mold properly and will resist all efforts to make good cheese.

6. When the clean break stage is achieved, it is time to cut the curd. Use a table knife to cut the curd in a series of quick movements, traveling from one side of the pot to the other. Next, cut in the other direction to produce small cubes. Cut quickly. The cubes should be about ½ inch when you are finished. Do not go smaller, or the curds will be too small and will be reduced to mush.

Fresh goat cheese is typically made in a round cylinder mold. The little discs of cheese are known as "crottins." After salting, allow to air dry 1 to 3 days to develop a little bit of a rind. The rind helps the cheese hold up in certain dishes. Gregory Gerber/Shutterstock.com

7. After the cutting is finished, allow the curds to rest for about 10 minutes. This allows them to heal.

8. Fill the molds to the top. Be gentle and scoop the curds; try not to break them into smaller pieces. Drain the molds on a rack. Whey will immediately begin to flow from the holes in the molds, and soon the cheese will begin to shrink.

9. Let the molds to sit for 2 to 4 hours, and then flip the cheese. (If at any time during this step the cheese does not appear solid enough to hold its form, wait 2 hours and try again.) To flip the cheese, remove it from the mold, turn it over in your hand, and put it back in the mold.

10. Allow the cheese to sit for about 12 hours at room temperature (approximately 70° F) over a rack or in your sink where it can continue to drain. After this draining period, the cheese will be ready to unmold. To remove the cheese from the molds, run a knife around the edge of the cheese and rap it on a hard surface. It will fall from the mold.

11. Dry the cheese for another 2 hours, and then move on to salting. Start with ½ teaspoon. Place the salt on wax paper, and pinch up a bit in your fingers; rub the salt on the top and sides of the cheese. Taste, and then adjust. Note: Salting is an acquired skill. It is easy to apply too much or too little. Err on the side of too little and then add more to taste.

12. If you would like to add herbs or seasonings, sprinkle the outside of the cheese with the desired herb now while it is still moist. Recommended herbs are fresh chives, herbes de Provence, coarsely ground black pepper, fresh mint or lemon balm, or fresh lemon thyme.

13. Place the cheese on a drying rack; a food dehydrator tray works quite well. The cheese should be dry in 4 to 6 hours. It will have developed a bit of a crust, which is fine. Wrap the dried cheese in specialty cheese paper, or wax paper as a substitute. Refrigerate. Due to the high moisture content, fresh cheese will keep only 15 days. The cheese will be at the peak of flavor when served at room temperature. Note: If you would like to add sundried tomatoes, olives, or other "toppings," allow the cheese to dry as above and add these extra ingredients on top of the cheese before serving.

Serving suggestion: Place a comb of fresh local honey in the center of a serving platter. Slice two or three firm Bosc pears and two goat cheese crottins. Surround the honeycomb with the pears and cheese, serve with crackers or toasted French bread.

Fresh Goat Cheese with pears and honey. This elegant yet simple dessert finishes a meal with flair. The honey offers a sweet contrast to the acidic bite of the cheese. The complexity of textures adds a bit of crunch, and the pear offers the perfect delivery for cheese and honey.
Olga Miltsova/Shutterstock.com

Chana (Paneer)

This delicious cheese recipe is from the collection of Nina Mukerjee Furstenau, author of *Biting through the Skin: An Indian Kitchen in America's Heartland* (www.ninafurstenau.com).

Ingredients and Equipment

1 gallon whole milk, as fresh as possible

½ to ¾ cup lemon juice

Cheesecloth

Yields 1¼ pounds

Paneer, the Indian cheese noted for a dense, dry texture and salty taste, is the perfect complement for spicy Indian dishes. Monkey Business Images/Shutterstock.com

1. In a heavy stockpot, bring the milk to a full boil, and when it starts to rise in the pot, add the lemon juice a little at a time until the milk separates. Strain the curd from the remaining liquid, and place it into cheesecloth.

2. After placing the cheese curds into the cheesecloth, squeeze out as much water as possible and place the cheesecloth and its contents on a clean baking sheet or the clean counter.

3. Fill the stockpot with water to make it heavy, and place it on top of the cheesecloth overnight.

4. In the morning, cut the dried Chana or Paneer into cubes.

Serve as plain cheese. Or, put a tablespoon of vegetable oil in a saucepan, and when hot, place one layer of cubes in and fry them until they are slightly brown; turn the cubes and repeat until all the pieces are golden. This cheese is wonderful in vegetable dishes like palak (spinach) paneer or in vegetarian curries.

Saag Paneer (Spinach cooked with cheese)

From the collection of Nina Mukerjee Furstenau.
Makes 4 servings

Ingredients

2 pounds fresh spinach, washed and stems trimmed (or two 10-ounce packages
   frozen spinach, thawed)
¼ cup ghee (or ¼ cup butter)
½ pound Chana (Paneer) Cheese (see recipe, page 53), cut into 1 x 2-inch rectangles
1 large onion, finely chopped
3 cloves of garlic, minced
1 to 3 green chilies, seeded and minced, optional
1 teaspoon freshly grated ginger
1 teaspoon Garam Masala, recipe follows (Garam Masala may also be purchased
   at eastern food stores.)
¾ cup plain yogurt
Salt to taste

1. Bring a large pot of water to a boil. Add spinach, and blanch for 1 minute until tender. Put the blanched spinach in a colander, and press it firmly with the back of a spoon to remove as much water as possible. Set aside.

2. Heat the ghee, or butter, in a heavy skillet over medium heat. Add the paneer pieces, and fry them for a couple of minutes, gently turning them until each piece is lightly browned on all sides. Remove the cheese from the skillet, and drain it on a paper towel for a couple of minutes before placing it on a plate.

3. Pour off any excess butter from the pan, leaving 1 to 2 tablespoons, and return the skillet to the heat. When the pan is hot, sauté the onions, garlic, and ginger for about 5 minutes until soft. Sprinkle the mixture with the garam masala and minced green chilies. Continue to stir 1 to 2 minutes. Mix in the spinach. Remove the pan from the heat and stir the yogurt into the spinach mixture. It should be somewhat thick. Gently mix in the fried paneer cubes, season with salt to taste, and serve with rice or Indian bread.

Spinach, paneer, peppers, and onions set the tone for the fragrant dish Saag Panner. Nina Mukerjee Furstenau states this was a popular dish in her family and one she now makes frequently for her own children. bonchan/Shutterstock.com

Garam Masala

Ingredients
2 tablespoons coriander seeds
2 tablespoons cumin seeds
1 teaspoon fennel seeds
2 teaspoons whole cloves
½ teaspoon black mustard seeds
1 tablespoon cardamom seeds, removed from their pods
1 tablespoon whole black peppercorns
2 dried red chilies, broken in pieces, seeds discarded
1 tablespoon turmeric

1. Place the whole spices (coriander, cumin, fennel, cloves, mustard, cardamom, peppercorns) and dried red chilies in a dry heavy-bottomed saucepan, and toast over medium heat for about 2 minutes. Shake the pan often to prevent burning.

2. In a clean coffee grinder, grind the toasted spices into a fine powder. Add the turmeric, and grind briefly again.

Use the spice blend immediately, or store it in a sealed jar for as long as 1 month. Makes about ½ cup.

Quick Cottage Cheese

Ingredients

1 gallon whole milk, pasteurized
⅛ teaspoon Mesophilic DVI MA Culture
⅛ teaspoon liquid rennet
¼ cup cool, nonchlorinated water
½ teaspoon salt
½ cup cream
Yields 1¼–1½ pounds

Equipment

Heavy cooking pot, large enough to
   hold 1 gallon of milk
Dairy thermometer
Slotted spoon
Ladle
Colander
Cheesecloth
Knife

Cottage cheese is a versatile product and a perfect way to use up excess milk. Wiktory/Shutterstock.com

1. Warm the milk to 86° F.

2. Remove the pot from heat, and stir in the culture.

3. Mix the rennet into the water, and add it to the milk. Let the mixture sit for 30 minutes, or until a clean break is achieved. Cut the curds into ½-inch cubes.

4. Heat the curds very slowly. The temperature should rise about 2 degrees every 5 minutes, bring the mixture up to 110° F. Hold the temperature there for about 20 minutes, stirring every few minutes to keep the curds from matting.

5. When the curds are firm, place them in a cloth-lined colander and let them drain for about 10 minutes.

6. Lift the cloth with the curds intact, and plunge the entire sack into cold water. Drain again until the curds stop dripping.

7. Place the curds in a bowl, and add a dressing of salt and cream.

Skimming cream from goat milk can be difficult due to the fact that the cream stays suspended within the milk. An easy way to skim the cream is to place the milk in a jar designed for sun tea, one with a spigot on the bottom. Pour the milk into the jar, and chill it. The majority of the milk that comes out of the spigot will be whole milk; however, the last bit, which will come out more slowly, will be the heavier cream. Save this part for the cottage cheese dressing.

Goat Butter

Goat butter is a delicacy. This creamy delight is more difficult to make than butter from cow milk due to the composition of the goat milk and the way the fat is suspended in the milk. However, if you use the methods of cream collection previously mentioned (a sun tea container with a spigot; see Quick Cottage Cheese opposite page), it is possible to make butter from goat milk. Look for antique molds online or at your favorite shop. These handcrafted wooden boxes with their dovetail joints are works of art. Sanitize the molds before use.

Ingredients

1 pint goat milk cream
Yields approximately 10 ounces

Equipment

Food processor (I find a food processor works better than a blender. The blender tends to liquefy the butter.)

Tools of the trade—an antique butter mold, butter paddle, and stamp. Butter is shaped into a block in the large wooden mold. The paddle assists the butter maker in pressing out excess buttermilk. Once formed, the butter is pushed from the block with the wooden plunger, then a wire is used to slice the blocks into quarters (sticks). A pat of butter is then stamped with the decorative handmade stamp.

1. Collect the cream, and allow the cream to ripen in the refrigerator for several days.

2. Put the cream into the processor bowl, and do four or five short bursts of mixing, leaving the machine on for a few seconds.

3. Turn the processor on full speed, and let it run. It may take 10 minutes or more to make butter. Listen for the processor to start to bog down as the butter comes. First, you will notice small pieces. With more processing, the pieces will come together to form a solid chunk.

4. Remove the (now) butter from the processor, and work it back and forth with old butter paddles from an antique store or wooden spoons. We want to get the buttermilk out or the butter will spoil quickly.

5. Line an antique wooden mold with wax paper for easy release. Once the moisture is worked out, put the butter in the mold; chill. Push the butter out.

Goat Butter will not be yellow but almost pure white. To make sticks, simply cut the butter into four sections with a knife or use a wire to cut down through the block. Typically the molds will produce a 1 pound block of butter. Use the leftover buttermilk to drink or in biscuit or pancake making.

30-Minute Fresh Mozzarella

This is an adaptation of Ricki Carroll's recipe. Ricki "The Cheese Queen" owns New England Cheesemaking Supply, a valuable resource for cheesemakers (www.cheesemaking.com).

Ingredients
1 gallon whole goat milk (pasteurized)
1½ teaspoons citric acid dissolved in ½ cup cool water
¼ teaspoon liquid rennet dissolved in ¼ cup cool, nonchlorinated water
½ to 1 teaspoon salt
Yields 1¼ pounds

Equipment
Heavy cooking pot, large enough to hold 1 gallon of milk
Dairy thermometer
Slotted spoon
Ladle
Colander
Cheesecloth
Rubber gloves
Heat resistant glass bowl
Microwave, recommended but optional

1. Before you begin, take the temperature of the milk. Why? This cheese moves very quickly, and 55° F happens sooner than you might think!

2. Place the milk in the pan, and heat it over medium heat until the milk reaches 55° F.

3. When the milk reaches 55° F, pour in the citric acid/water mix and stir for 15 seconds. Continue to heat the milk over medium heat until it reaches 90° F, stirring once or twice.

4. When the milk reaches 90° F, add the rennet/water mix and stir gently for about 30 seconds. Then, stop stirring and let the curd form.

5. Continue to heat the milk until it reaches 105° F. Turn off the heat, and remove the pan from the burner. Wait 5 minutes and the curd will be pulling away from the side of the pan. The curd will look like thick yogurt and have a bit of a shine.

6. Scoop out the curds with a spoon, and put them into the cheesecloth-lined colander. There will be a lot of liquid.

7. Roll the curd back and forth in the colander while keeping it over a bucket. This will help to form the curd and drain the whey.

8. Transfer the curd to the glass bowl, and drain off more whey.

9. Bring your bowl to the microwave, and microwave the curd for 1 minute. Gently fold the cheese over with your hand or spoon. It will be hot! Wear rubber gloves. Drain off more whey.

10. Microwave the curd again for 35 seconds, and add salt. Knead the curd again. Drain.

11. Microwave again for 35 seconds, and knead the curd. Drain.

12. Knead the curd gently until it is smooth and elastic. When the cheese stretches like taffy, it is done.

When the cheese is smooth and shiny, roll it into a ball and eat it while it is still warm. If you want to save the cheese to eat later, place it in a bowl of ice water for 30 minutes to bring the inside temperature down rapidly. Even with this treatment, it is best to eat the cheese while it is still fresh.

If you do not have a microwave, heat the reserved whey to at least 175° F. Add ¼ cup noniodized salt to the whey. Shape the cheese into one or more balls, put them in a ladle or strainer, and dip them into the hot whey for several seconds. Knead the curds between each dip, and repeat this process several times until the curd is smooth and pliable. Make sure you wear rubber gloves! Tip: Keep a bowl of ice water nearby, and when your hands get too hot, put them in ice water, chill, and then resume cheesemaking.

Whey Ricotta

Ingredients

2 gallons fresh whey
2 cups whole goat milk, pasteurized
¼ cup cider vinegar
½ teaspoon noniodized salt
Heavy cream, optional
Yields ½ cup

Equipment

Heavy cooking pot, large enough to hold 2.5 gallons
Dairy thermometer
Slotted spoon
Ladle
Colander
Cheesecloth

1. Pour the whey into a large pot, and add the milk. Heat the milk whey to 190° F.

2. While stirring, turn off the heat and add the vinegar. You will notice tiny white particles.

3. Carefully ladle the curds into a colander lined with cheesecloth.

4. Allow the curds to drain for several hours. Add salt and a small amount of cream, if desired.

Cover and refrigerate. Use within 1 week.

I equate cheesemaking with bread baking. If you have ever worked much with yeast, you will know that on certain days the gods appear to be working against you. No matter what, the bread refuses to rise. Cheesemaking can be like that, too. No matter what you do, the milk may refuse to set. What can cause this? There are a variety of things to consider.

- Was the milk fresh?
- Could there have been bleach or sanitizer left in any of the containers? Chlorine can interfere with rennet.
- Is the goat in late lactation? If so, the actual composition of the milk can change. If this is the case, at this point I would introduce some calcium chloride to enhance the curd structure.
- Has the weather been hot and dry? This, too, can affect the milk and cause changes that make cheesemaking more difficult. Add the calcium chloride, and try again.
- How old are the cultures? Most keep 1 year if kept in the freezer. Rennet lasts about 6 to 9 months in the refrigerator; 12 months for frozen tablets.

If none of those questions provide an answer, delve deeper into the process. Once in a great while, particularly during hot weather, the cheese may actually blow in the mold. By this, I am referring to the process in which cheese normally shrinks in the mold and becomes solid. If there is a bacterial problem in the milk—caused by unsanitary milking conditions, high bacteria counts, or health problems with the goat—the cheese will become springy and have an off smell. There will be small holes throughout the cheese. If this happens, discard the cheese. Check the goats for signs of mastitis, change your milking routine, and sanitize your utensils and milking parlor. Chill the milk immediately after milking as previously mentioned. Room temperatures can also play a role when draining cheese. If the room is too hot, the whey can sour and spoil the cheese; 70 to 74° F is optimal. Hot, humid conditions present a challenge to the goats and the cheesemaker.

Newly freshened goats will produce colostrum, the first milk that is necessary for the kids. It is full of vitamins and antibodies that are crucial to the kids' survival. However, if any of the colostrum is in the milk intended for cheesemaking, this will also cause problems with the setting of the curd. Allow 3 to 4 days of milking before saving the milk for cheesemaking.

Over time, experience will help you to anticipate and recognize problems early in the process. Cheesemaking can be trial and error until you get it down, then you will have it forever and feel like you have conquered the world! Cheesemakers are all alchemists at heart.

Yogurt

Yogurt made from goat milk is tangy and pleasantly acidic. Often, those who are allergic to cow milk can tolerate goat-milk products. This recipe comes from Jennifer Bice of Redwood Hill Farm and Creamery.

Ingredients

1 gallon fresh goat milk, unpasteurized

1 tablespoon plain yogurt with active cultures, or 1 packet freeze-dried culture containing lactobacillus

Cassava root, as thickener, optional

Yields 64 ounces

Equipment

Heavy cooking pot, large enough to hold 1 gallon of milk

Dairy thermometer

Slotted spoon

Ladle

Sterilized jars

Incubator (oven, heating pad, crock pot, etc.)

1. Heat the milk to 108° F.

2. Add the plain yogurt or freeze-dried cultures. Make sure to use yogurt from a new cup and a clean spoon to add it.

3. Incubate the milk mixture at 104 to 108° F using a home yogurt maker or an incubation device of your own. Some people use a heating pad wrapped around a jar, put the jar in the oven on low, or place the jar in a crock pot. Whatever you decide to use, experiment with water and a thermometer before you actually make the yogurt, and make sure you can hold the milk at the required temperature.

4. Incubate the milk for 6 to 8 hours, depending on your taste.

5. When the milk is finished incubating, chill the yogurt before eating it, being careful not to agitate or move the yogurt much until it is well chilled.

Goat-milk yogurt will not get as thick as cow milk yogurt. Many commercial cow milk yogurts add powdered milk as a thickener. You can also use a small amount of tapioca, which is a natural thickener from the cassava root.

Fresh yogurt is so easy to make. Follow Jennifer Bice's recipe from Redwood Hill Farm. Serve with honey or maple syrup if you like it sweet. Drain unsweetened yogurt in a yogurt funnel or a funnel lined with cheesecloth to make a tangy cheese spread.

Feta

In this cheese, yogurt, with live active cultures takes the place of the mesophilic culture.

Ingredients

1 gallon goat milk, pasteurized
½ teaspoon liquid rennet dissolved in ¼ cup nonchlorinated water
1 tablespoon plain yogurt
Yields 1 pound

Equipment

Heavy cooking pot, large enough to hold 1 gallon
Dairy thermometer
Slotted spoon
Ladle
Whisk
Colander
Cheesecloth
String for hanging cheese

1. Pour the milk into the pot, and heat it to 86° F. Remove the pot from heat.

2. Add the yogurt to the pot, and blend it lightly into the milk with a whisk.

3. Stir ½ teaspoon liquid rennet into ¼ cup nonchlorinated water, and add the rennet/water to the milk.

4. Let the mixture rest 12 hours, covered with a tea towel, at room temperature (68 to 72° F).

5. After 12 hours, check the curds. You should see a clear distinction between the solid curd and the liquid whey. If not, wait 2 to 3 more hours.

6. Cut the curd into ½-inch pieces, making sure to slice all the way through to the bottom of the pot. It may be necessary to make some diagonal cuts to avoid long curd "noodles." All the curds should be approximately the same size. This uniformity helps them go back together again.

7. Let the curds rest for 20 minutes, stirring every 5 minutes.

8. Using a colander lined with cheesecloth, drain the curds from the whey. (Save the whey for the brine.) Ladle the curds into the cloth, and the whey will naturally separate.

9. Bring the corners of the cheesecloth into a bag, and hang it to drain for 3 to 4 hours. Then reposition the cheese in the cloth to give it a nice round shape. Let it continue to hang for 24 hours.

Feta, a popular cheese originating in Greece, picks up salt as it ripens in brine and is perfect for complementing Greek dishes and Middle Eastern foods. Feta is traditional on gyros, which are served on a pita with yogurt tzatziki sauce. In the Middle East, feta may be simply known as "white cheese."
Karl Allgaeuer/Shutterstock.com

10. Take the cheese from the cloth, and cut it into 2-inch cubes.

11. Prepare a brine of ⅛ cup noniodized salt to 4 cups of whey. Add room temperature water if needed to reach 4 cups. Combine the whey (and water, if needed) and salt in a glass container with a wide mouth.

12. Place the feta cubes into the brine, then refrigerate for 7 days.

To use, remove the cubes from the brine, rinse, and crumble. Serve immediately. Leave the remaining cheese in cubes, pat them dry, and store them in a zip-top bag. This cheese will keep for up to a year due to the high salt content, as long as it is not too moist.

Feta in Olive Oil with Sun-Dried Tomatoes

Ingredients
Feta cheese
Onion, sliced
Garlic cloves
Sun-dried tomatoes
Fresh whole basil leaves
Olive oil
Yields four 8-ounce jars

Equipment
Pint jar

1. Make the recipe for Feta as on page 62.

2. Brine the cheese for 1 month, then remove 8 to 12 cubes.

3. Place the cubes in a pint jar. Add onion slices, garlic cloves, sun-dried tomatoes, and fresh basil leaves. Arrange the basil leaves around the outside of the jar so they look attractive; layer the cheese with the sliced onions, garlic, and tomatoes.

4. Fill the jar with olive oil, and refrigerate it for a few days for the flavors to blend.

Serve at room temperature. The oil picks up the flavors and makes an excellent salad dressing or dip for warm French bread.

This recipe can accommodate other vegetables, such as zucihini and fresh peppers.

Digiriic/
Shutterstock.com

Queso Fresco

Ingredients

1 gallon goat milk, pasteurized

¼ cup white vinegar

1 cup plain yogurt with live active cultures, to take the place of the Mesophilic culture

⅛ teaspoon calf/mild picante lipase, optional

2 to 3 teaspoons noniodized salt

Yields 1 pound

Equipment

Heavy cooking pot, large enough to hold 1 gallon

Dairy thermometer

Slotted spoon

Ladle

Whisk

Colander

Cheesecloth

String for hanging cheese

When made from goat milk, queso fresco ("fresh cheese") will be very white and have a crumbly texture, great for topping enchiladas or other Spanish dishes. This cheese also has a nice melting quality for dishes such as stuffed poblano peppers.

1. Pour the milk into the pot, and heat it to 88° F.

2. Add the yogurt, and blend it in, stirring lightly. If adding lipase, add it now.

3. Let the milk/yogurt rest for 4 to 6 hours. Cover the pot to keep the heat in for as long as possible. This waiting period allows the bacteria in the yogurt to multiply and ripens the milk.

4. Put the pot back on the heat, and stir gently until the temperature reaches 185° F.

5. Remove the pot from heat, and pour in the vinegar in a steady stream. Wait 10 to 15 minutes. At that time you should see distinct curds and whey.

6. Line a colander with a flour sack or dish towel, and ladle in the curds and whey.

7. Gather the corners of the towel, and bring them together to form a bag. Tie the bag, and hang it to drain for 1 hour.

8. Take the bag down, and break up the curds; add salt.

This is a crumbly cheese, intended for use on top of beans, burritos, tacos, and so on. It is somewhat bland and is often served with a spicy dish to absorb some of the heat. The lipase gives it additional flavor.

Yogurt Cheese (Lebneh)

Lebneh yogurt cheese is widely used in the Middle East and Greece—a fact that results in several spellings of its name, including *lebanah* and *labanah*. This cheese is the same consistency of cream cheese. It is also easy to make. Simply allow the liquid to drain from the solids. Yogurt funnels are available for this very purpose. Source: Jennifer Bice of Redwood Hill Farm and Creamery

Ingredients
2 cups plain yogurt
Yields 8 ounces

Equipment
Yogurt funnels or a colander lined with cheesecloth

1. Place 2 cups of plain goat yogurt in a colander lined with three layers of moistened cheesecloth.

2. Bring the corners of the cheesecloth together to form a bag, which can then be drained over the sink. Let the yogurt drain for 8 to 16 hours.

3. Stir occasionally, scraping the cheese away from the cheesecloth to allow better draining. The longer the yogurt drains, the thicker and more tart the yogurt cheese will be.

4. When the desired consistency is reached, add herbs, caraway seeds or fresh black pepper as desired. Salt is optional.

Refrigerate and use within 2 weeks.

Kochkaese

A classic German cheese, the name translates to "cooked cheese." I live in a German settlement, and countless neighbors have told me stories about their grandmothers making this cheese.

Ingredients
2 cups freshly made chevre (see Chevre recipe, page 46)
½ cup goat milk
¼ cup butter
½ teaspoon baking soda
Salt
Yields 10 ounces

Equipment
Heavy skillet (Cast iron is good.)
Wooden spoon, for stirring

1. Place the chevre in a large, heavy skillet, and add the milk, butter, baking soda, and a heavy sprinkling of salt.

2. Place the skillet over low heat, and stir constantly until the milk and butter are worked in. Cook about 20 minutes, stirring continuously.

3. Continue to cook and stir over low heat until the mixture thickens and does not look shiny. The result will be a thick, spreadable cheese. Add black pepper or caraway seeds, if desired.

Store the cheese in the refrigerator in a covered container. This spread will keep for about 2 weeks.

Kochkaese, a simple cheese that is a staple on the German table, translates to "cooked cheese." Fresh curd is cooked with butter to make a rich, spreadable, tangy cheese. The perfect pairing is rye bread and a German-style lager.

Sweet Potato Salad

This recipe highlights Fresh Goat Cheese as an ingredient.

Ingredients

4 medium-sized sweet potatoes, peeled and cooked
4 green onions sliced
1 celery stalk, chopped fine
1 Gala apple, chopped into medium-sized pieces
½ cup Vegenaise or mayonnaise
½ cup Fresh Goat Cheese (see page 51), cut it into ½ pieces
Fresh chives for garnish
Salt and pepper
Makes six ½-cup servings

1. In a saucepan, cover the potatoes with water and boil them until they are tender but firm. Remove them from the water, and cut them into ½-inch pieces.

2. Add the sliced green onions, the chopped celery, and the apple. Toss with Vegenaise or mayo.

3. Add the goat cheese and stir lightly, just enough to distribute the cheese through the salad.

4. Add salt and pepper to taste.

Refrigerate until chilled; garnish with freshly snipped chives

Sweet potatoes are often on the menu at my house. This vitamin-packed root vegetable is often overlooked as a tasty side dish. Whether baked, boiled, mashed with carrots and parsnips, or served in salad, sweet potatoes are versatile and unexpected. I was experimenting one day and came up with this sweet potato salad recipe. The addition of goat cheese was one of those happy accidents.

Stuffed Shells

For a new twist, use goat-milk ricotta or chevre and fresh mozzarella to make stuffed shells. This recipe is a family favorite

Ingredients

1 box jumbo pasta shells, uncooked
1 jar prepared marinara sauce, traditional
2 eggs
3½ cups Whey Ricotta (see page 59), or substitute 3¼ cups Chevre (see pages 46–47) blended with ½ cup goat milk
4 cups shredded 30-Minute Fresh Mozzarella Cheese (see pages 58–59), divided
1½ cups grated Parmesan cheese, divided
1 tablespoon dried parsley
Makes 6 to 8 servings

1. Preheat the oven to 350° F.

2. Cook the shells according to package directions; drain.

3. Beat the eggs in a large bowl. Stir in the ricotta or chevre with the goat milk, 3 cups of the mozzarella, 1 cup of Parmesan, and the parsley.

4. Fill each cooked shell with the ricotta mixture. Arrange the filled shells in a 15x10x2-inch baking dish. Top with remaining marinara sauce. Add the remaining Parmesan on top.

5. Bake, covered with foil, until bubbly, about 45 minutes. Uncover. Cook until the cheese is melted, about 5 minutes. Let the shells stand 5 minutes before serving.

Truffles

Ingredients
6 ounces bittersweet chocolate
6 ounces chevre
¼ cup powdered sugar
½ teaspoon pure vanilla extract
20 pecans, optional
Raspberry liquor
¼ cup cocoa powder
Makes about 20 truffles

Equipment
Double boiler
Metal spoon
Whisk
Mixing bowl
Waxed paper
Cookie sheet
Injector, optional

1. Set up the double boiler, and melt the bittersweet chocolate. (Note: Use a good chocolate; the better the chocolate the better the truffle!) Stir the chocolate until it is thoroughly melted and smooth. Set aside.

2. In a small mixing bowl, whisk together the chevre, sugar, and vanilla until light and fluffy. Add the melted chocolate, and whisk until well blended. Cover and chill for 1 hour or until firm.

3. Using a tablespoon, drop rounded balls of the mixture into the cocoa powder. Roll to coat them with cocoa. Place them on a wax paper–lined cookie sheet; chill. Variations: Shape the chocolate mix around a pecan, then roll the truffle in the cocoa powder and chill, or shape the truffle into a firm ball, roll in the cocoa powder, and chill.

4. After thoroughly chilled, inject with ½ teaspoon raspberry liquor and chill again until serving.

Omelets with Fresh Goat Cheese

There is nothing finer than sitting down to a farm fresh meal. Omelets highlight the "hen fruit," fresh tomatoes, cilantro, and green onions all from the garden, and of course, goat cheese, the barnyard complement.

Yields 1 large omelet

Ingredients, per omelet
Pat of butter
3 large fresh eggs, whisked
   until frothy
4 slices thinly sliced Fresh
   Goat Cheese
1 ripe tomato, chopped
Several springs of cilantro,
   chopped fine
2 green onions, chopped
Juice of ½ lime
Salt and pepper to taste

1. Heat a small sauté or omelet pan, and melt a pat of butter. Move the pan to evenly distribute the butter.

2. Add the whisked eggs; cook until the eggs begin to set. Then, using a spatula, move the eggs, allowing the uncooked portion to flow to the edge of the pan.

3. As the eggs firm, add the chopped tomato mixed with lime juice, cilantro, and onions. Top with fresh cheese.

Perfect for breakfast, lunch, or dinner, omelets are always a favorite. Fresh goat cheese goes perfectly with eggs. Add fresh pico de gallo and warm tortillas for a quick and filling meal. Garnish with pepper jelly for an unexpected punch.

4. Flip the top over the omelet, and allow it to remain on the heat 1 minute or until nicely browned.

Serve with fresh tortillas, Pepper Jelly (recipe follows), black beans, and sour cream. One omelet serves one hungry person.

Pepper Jelly

Ingredients:
2 to 3 jalapeño peppers, seeded and chopped (Wear gloves!)
2 cups finely chopped red bell peppers
2 cups finely chopped green peppers
5 cups sugar
1 cup apple cider vinegar
1 (1.75 ounce) package powdered pectin
Yields eight 4-ounce jars

Equipment
Nonaluminum saucepan
Wooden stirring spoon
Sterilized jars
Canner

1. Combine the peppers and vinegar in a large nonaluminum saucepan. Add the vinegar and fruit pectin; mix well.

2. Bring the mixture to a full rolling boil, and add the sugar, a cup at a time. Bring the mixture back to a full rolling boil, and boil for 1 minute. Stir constantly.

3. Skim off any foam that forms, and remove the saucepan from the heat. Remove the pepper pieces, if desired.

4. Pour the hot jelly into sterilized jars. Leave a ¼-inch head space, and top the jar with a lid and ring. To process the jars in a water bath canner: Place the jars in a canner rack and lower it slowly into the canner filled with enough hot water to cover the tops of the jars. Bring the water to a boil. Boil for 5 full minutes at a full boil. Remove the jars with a jar lifter, and place them on a towel to cool.

One day a year is set aside to make pepper jelly. A pantry staple, this jelly is surprisingly complementary to cheese. Whether served with fresh chevre on a good cracker or alongside fresh goat cheese with rustic bread, the slight fire of the jelly combines with the sweetness of jelly to make an unexpected treat. Make sure to wear gloves when working with hot peppers.

Goat Cheese Dip

Ingredients
4 ounces freshly made Chevre
½ cup sour cream
1 tablespoon lemon juice
⅛ cup chopped chives
Salt to taste
Yields 8 ounces

1. Mix the first four ingredients, then fold in the chives. Add salt to taste.
2. Place in a ramekin, and let chill.

Top with fresh chive blossoms, and serve with crackers.

Goat-Milk Ice Cream

This recipe requires only two ingredients and a small ice cream freezer!

Ingredients
2 cups goat milk, pasteurized
½ cup pure maple syrup
Makes six ½-cup servings

1. Stir the milk and syrup together.

2. Place the mixture in a frozen canister.

3. Freeze and enjoy.

Honey Ice Cream

Ingredients
2 cups goat milk, pasteurized
½ cup honey
Makes six ½-cup servings

1. Blend the milk and honey together.
2. Place the mixture in a frozen canister, and blend until frozen.

Cajeta

Cajeta is a caramel spread that can only be made from goat milk. It is popular in Mexico and is used in desserts or with flan. Try it on top of ice cream!

Ingredients
2 cups goat milk
2 cups sugar
⅛ teaspoon pure vanilla extract
½ teaspoon baking soda dissolved in 1 tablespoon water
Yields 8 ounces

Equipment
Heavy cooking pot, large enough to hold 1 gallon of milk
Dairy thermometer
Wooden spoon
Ladle
Container, for storage

1. Combine the milk, sugar, and vanilla in the pot.
2. Place the pot over medium heat; stir until well blended and the sugar is dissolved.
3. Remove the mixture from the heat, and stir in the baking soda and water. This addition will cause the milk/sugar to bubble. Wait until the bubbling stops, and then return the pot to medium heat.
4. Bring the liquid to a simmer, and stir continuously. It can take as much as 1 hour or more for the caramel to form. Watch this pot carefully; the liquid is easy to scorch and the caramel forms quickly. We are not making brittle! Look for a nice smooth consistency, like a pourable caramel topping. The color of the milk will change to a distinctively caramel color.

   When this stage is reached, remove the pot from the heat and pour the caramel sauce into a heat resistant container or glass jars. Enjoy some while warm! Allow the remainder to cool, and refrigerate it. Serve with ice cream, brownies, or apples, or stir some into a fresh glass of goat milk.

Goat-Milk Ice Cream is surely the food of the gods! This is a simple recipe of goat milk and real maple syrup. Freeze in a tabletop ice cream freezer, and this sweet treat will be ready in about 30 minutes. Top with Cajeta, a Mexican-style caramel. David P. Smith/Shutterstock.com

Goats for Meat

It should be stated that any type of healthy, disease-free goat may be consumed for meat. In fact, dairy and fiber goat breeders often consume or market their excess males as goat meat. This is a perfectly acceptable practice, and young goat makes for some fine eating.

If you are familiar with the cattle industry, you know that there are certain breeds that are best for milk and others that are best for meat. This does not prevent a farmer from turning a Holstein into a "hamburger cow." It is the same with goats. It all comes down to the quality of the meat, the texture, tenderness, and overall flavor. If producing goats for the meat market is the primary goal of your enterprise, then consider the following breeds.

Spanish meat goats. These descendants of goats were brought into the United States by early New England settlers. They migrated south and probably interbred with goats brought into Texas and Mexico by early Spanish settlers. These goats are rangy, smaller than some of the other meat-specific breeds, but extremely hardy because of their adaptation to their rugged living conditions.

Boer goats. This South African breed has become a very popular meat goat in the United States. The distinctive white body with reddish brown to black head, ears, and markings makes these animals easily identifiable. The

attributes of this breed include stature (stocky and muscular), good fertility rates, and high rates of gain and growth.

Tennessee meat goats. Also known as "myotonic" or "fainting" goats, this breed has the odd characteristic in which the animal actually faints when startled. Their muscles lock up and over they go, taking a few moments to collect themselves and to return to their original state. This happening can be somewhat comical and unexpected. One goat breeder shared with me that she went to the garden wearing bright yellow slacks. Her attire caused a mass fainting of

Boer goat colors range from blond to red to dark brownish/black. Some have splashes of color on their torso.

Note the ears, body structure, and conformation of this healthy Boer kid. Boers have a stockier bodies and shorter legs than dairy goats. Their coats are typically white with patches of color and a dark head.

the goats in a nearby pen. What fashion critics! The good news is that the goats quickly recover from their fainting spells and they exhibit a good rate of gain. Due to the fact their back legs stiffen when they faint, the muscle growth is enhanced, making for nice cuts of meat.

Kiko goats. Kikos originated in New Zealand and are the result of crossing feral does with Saanen and Nubian bucks. Kikos are a relatively new breed, only recognized since the 1980s. *Kiko* is the Maori word for "flesh" or "meat." Known for their overall hardiness and low input costs, Kikos have gained popularity since their introduction to the goat world.

❦ The Meat Market ❦

Goats are widely consumed all over the world. As the population of the United States changes each day, the demand for goat meat is higher than ever before. Cultural celebrations, personal tastes, and regional cuisines have served to grow this market.

Due to this demand, the number of USDA-inspected processing facilities available has grown substantially. If you are considering a meat goat operation, check with your state Department of Agriculture to locate goat-processing facilities in your area. The processor will often purchase the animal on a live weight basis. Consider the cost of hauling or contracting with a hauler as you work out your overall budget. Seasonality is also a factor on the payment rate. Check with the processing facility to see when the highest rates are paid and adjust your breeding program to maximize your payment.

High demand drives up the price of goat meat, and these demands are often associated with religious holidays and cultural celebrations. This list includes the holiday, approximate date, and type of goat desired.

Holiday	Date	Type of Goat
Epiphany	January	milk-fed kids (18 pounds or more live weight)
Western Roman Easter	check calendar	fleshy milk-fed kids 3 months old or younger, 20 – 50 pounds (30 pounds considered optimum by most buyers)
Eastern Orthodox Easter	check calendar	slighter larger milk-fed kids, 35 pounds optimum
Cinco de Mayo	May 5	20 to 35 pounds live weight, milk-fed kids for cabrito; larger for seco de chivo
Independence Day	July 4	kids or young bucks, does, and wethers 1 year or under in age, suitable for barbecue, cabrito
Start of Ramadan	check calendar	uncastrated males or female kids with all their milkteeth (under 12 months of age); optimum weight 60 pounds (45 – 120 pounds accepted by different buyers)
Eid ul-Fitr	check calendar	Same animals as for Ramadan
Eidul-adha (Festival of Sacrifice)	check calendar	blemish-free yearlings (not castrated)

Making Alterations

If you intend to raise goats for the meat market, check with local buyers before you alter the animals in any way. Do NOT disbud or castrate until you check and see what the meat market desires. Often, buyers prefer the animal to be unaltered to adhere to certain religious practices. Meat and fiber goats are usually left with their horns intact.

✤ Maintaining a Healthy Herd ✤

Meat goats should be maintained with parasite management, scheduled vaccines, and overall health maintenance. Anecdotal reports and personal experiences have found that the does of the Boer breed frequently abandon their offspring. Penning the animal prior to birth can encourage the bond between mother and newborn. Many goat breeders simply allow these animals to live on open range, which makes penning unrealistic. Be prepared to make some decisions about bottle feeding, grafting the kid to a willing mother, or accepting loss. Kikos, on the other hand, are known as devoted mothers. Consider your breed carefully before making selections.

When new animals are introduced to your herd, the animals should be placed in a holding pen for 2 to 3 weeks. This time will allow for the discovery of any illnesses and allow the animals a gentle introduction to their new surroundings. It is best if the new animals can see the established herd and vice versa.

Bring the new goats up to date on immunizations, worm them, and watch them for any signs of illness. Running a fecal sample prior to worming is advised. This test will reveal the parasite type and load, allowing for accurate dosing of wormer. Keep an eye on the new animals when they are released to the herd. Just like a new kid on the first day of school, there will be a lot of curiosity from the other goats. Make sure the new additions are not bullied or harmed by overzealous pasture mates.

Speaking of pasture, as previously stated, goats are browsers, which means they prefer forage and brush to grass. When running fence, keep this in mind. If at all possible provide the goats with wooded or brush-filled areas in addition to grass. A rotational system can be applicable to a goat operation. Envision a wheel, and divide your property into spokes (sections of pasture) that come to a central section in the center of the wheel. In this central area provide a fresh water tank and perhaps a small pen to work the animals in. Close off sections of the wheel and allow those to grow, rotating the animals frequently to new pasture and browse. This method of grazing also cuts down on parasite infestations.

Fencing

Give fencing a great deal of thought before you get started with goats. Fencing will keep goat rearing an enjoyable experience—or create a daily problem. This choice is up to the landowner, and the proper installation must be underway from day one. Goats can jump without getting a run to clear a fence line; they appear to simply will themselves over, and the deed is done. Once freedom is tasted, the problem becomes established, and soon other animals will follow. My advice to any

Life in the meadow. Early spring growth provides perfect fodder for goats.

new goat owner would be to have your fencing installed before bringing home the first animals. If you put the cart before the horse (or the goat before the fence) problems are sure to follow.

A number of methods will efficiently and safely contain animals, including electrified poly netting, traditional electric fencing, and woven wire. Electrified poly netting is a fairly recently developed product that provides the means for containment without the super structure of a traditional fence. The poly netting is set up by setting metal fence posts into the ground, more or less creating an instant pen. This method is great for those who wish to move their animals to fresh pasture frequently. It is best used on fairly open pastures, as it can be difficult to run through a wooded section. The fence can be electrified with a traditional electric (110 volt) fence charger or a solar charger. Solar chargers have evolved over time and are efficient means of electrifying a fence. Harnessing the sun's energy also allows for electrified fencing in remote areas where no power is available. Research is currently being done on a system, similar to that which is used for dog containment, in which the electrified fence is buried beneath the ground and the goat wears a device that initiates a shock when it attempts to cross the barrier.

A traditional electric fence can also be utilized as a more permanent type of fencing. Five strands of aluminum wire conduct electricity via a charger plugged into a 110-volt outlet. Frequent maintenance of grass or weeds that may weaken the charge or short out the fence is required. It won't take long for the animals to notice the weak charge and head out under the fence. You know the grass is always greener! Heavily coated animals also may not respond to an electric charge due to the fact that their coat will insulate them from the shock—to the point they will consider it worth the risk to duck under. Of course, caution should always be used around electrified fencing, be it traditional wire or netting. Make sure small children and pets do not become engaged with the fence or tangled in the netting. It is always better to err on the side of safety. This fence is run in a traditional manner with corner posts and gate posts of cedar or hedge (Osage orange). T posts are then driven into the ground and aluminum wire is run along the T posts and clipped with plastic insulators.

The charge running through the fence should not be "hot" enough to harm an animal—merely to shock it and send it in another direction. Always be cautious of

Fencing is hard work. Rather than break a fence with a gate, some goatkeepers add a set of stairs for the goats to climb. This gives them a place to play and a means to go from one pasture to another.

young children around electric fences, particularly if you have young visitors who are not farm kids.

For a large permanent area, woven wire is the most expensive but also the most stationary and reliable. This fence should be run in a traditional manner with wooden end and gate posts, as well as T posts set into the ground. Care should be taken to ensure there are no spaces under the woven wire where a curious goat can (and will) nose its way through. Often those small gaps can be filled in with dirt and rock or a fir tree cut down and stuffed in the hole. Since most land isn't perfectly flat, watching for spaces is an important part of keeping your goats where you want them.

Goats are attracted to sweet things, salty things, and color. If the goats escape from their pens or pastures, they will head for flowers, laundry on the clothesline, and the highest place they can find. They will tap dance on the new car and potentially eat the upholstery if the opportunity is given. In general, people who say they hate goats have had a few and didn't have proper fencing. I've known a few animals who have consumed wicker porch chairs, not to mention ferns, gardens, and any other substance they considered to be to their taste. Remember, Rule #1: Good fences will be worth the cost in the long run. Goats can also run quickly, and they do resist being caught. They seem to have an extraordinary sense of man vs. beast and will attempt to outwit their human handler at most opportunities. Even the most docile animal can become excited at the opportunity to free range and take full advantage of the opportunity at hand.

Human Health Benefits

I have often read that people who have animals in their life usually have lower blood pressure levels. If you do not invest in good quality fencing, this will certainly not be the case with goats. Human blood pressure levels have been known to go through the roof after a romp through the hay field, chasing a goat that always seems to be inches from your grasp. Good fences not only make good neighbors, but they also keep the goats in.

Compadres. There is a pecking order in place here. The dog is the leader of the herd and watches over the goats night and day. Second in command is the lead doe, who guards the door and is always first to the milking stanchion. Goats usually line up in the same order each day when it comes to milking time.

Guard Animals

In my own operation, I always pulled the animals into a central closed pen at night. This precaution limited predator loss. Some producers may not have this option and may want to consider guard animals.

Popular animals for guarding goats are dogs such as Pyrenees, Anatolian shepherds, Australian shepherds, and other breeds. Additionally, llamas and donkeys fit well into the role of guardian. I also ran a donkey in with my herd, even when I penned them at night, and when I took care of 300-plus head, I used llamas as guard animals for goats on an open range. In the middle of coyote country, there was never a recorded loss in over 4 years with the llamas on duty.

Before you decide on a guard animal, talk with other goat producers to learn the pros and cons of each animal. Sometimes the guardian can be too aggressive toward its own herd. I've seen this with dogs and donkeys. Donkeys do appear to be amazing in that I believe they "count" their herd, and if

an animal is missing, they become very upset, letting the farmer know something is amiss. I also found that when bringing home a new animal that I had to "introduce" the new herd member to the donkey to avoid problems. If I missed this step, the donkey assumed this animal was not part of its herd and therefore should not be in the pen.

Guard animals, such as this Great Pyrenees, provide a valuable service on the farm. Goats are largely without defense, and predator loss is common. This fellow takes his job quite seriously and is the guardian of the herd. Alain Lauga/ Shutterstock.com

Donkeys are also often employed as guardians. My donkey, Tarzan, kept a close eye on his charges, and any new animal had to be properly introduced. A gentle giant, he would often herd misplaced kids back to their mother. His bray welcomed the sun each day.

I observed my donkey being quite aggressive to the goats at times—to the point that I put a board across a section of fence so only the goats would have access and he had to remain outside of this area. Arguments centered upon the hay bale, so feeding goats in the smaller pen and having a separate bale for donkey helped some. However, the goats would still torment him, and he did resent it.

On the other hand, one day I witnessed him herding two newborn kids between his feet, heading them back up the hill to their mother. He was extremely gentle and knew his steps had to be well placed. A few days later I glanced out, only to see one of those kids had jumped off a hay bale and was standing on the donkey's back. What a picture! The donkey loved it, and the kid danced around apparently scratching places that were hard to reach. There is never a dull moment with animals!

Water and Food

Periodically wash down water troughs to avoid algae growth and off-tasting water. If the goats do not consume enough water, health problems will occur. Always make sure the water supply is fresh and clean.

So what should a goat eat? Goats prefer brush and forage. They will eat grass, but it is not their food of choice. A good mixture of both is an acceptable combination. Pelleted feed is recommended as an additional source of nutrition particularly if the animal is intended to produce meat, fiber, or milk or is pregnant.

In the winter, supplemental hay will be required in areas where grass and forage are no longer available. Mixed grass hay is a good choice. A third cutting will often yield more coarse vegetation, which the animals eat readily. Watch for signs of overconsumption, and restrict the amount introduced for the first few days to avoid gorging. A key hole feeder is recommended.

Commercial goat rations are readily available. Follow the manufacturer's instructions for quantity of food. Remember, there will be weak and strong members of the herd. Make sure the poor doers get to the trough, or they will become weaker by the day. At times, separation of the small and weak is advisable.

❦ Routine Care ❦

Routine care begins in the barn. If animals are provided with enclosed housing, the barn is the place where disease, unsanitary conditions, and other problems can begin. Routine cleaning, manure disposal, and exposure to fresh air will greatly lower the potential for problems. Remove soiled hay and waste on a regular basis. For the most part, the goats will be healthier if access to the barn is limited. Disease can become established in enclosed places and remain there for years. That being said, a barn can be an asset in the freezing winter months.

Many goat producers build bunks within the barn. This keeps the animals from drafts

Llamas are another choice for guarding a herd. They are often more aloof but take their role of guardian very seriously. Their main defenses are their hooves and off-putting spit! Llamas provide beautiful fiber as a bonus for those who spin and weave. Imagine a mohair/llama blend garment. Priceless.

and cold that seeps up through the floor. Some farmers prefer a dirt floor, others concrete. There are pros and cons to both choices. Concrete can be cold; dirt floors are hard to keep clean. A layer of straw will help in both situations.

Other than during kidding season or times of weather extremes, the best place for the goats is out in the pasture. Canopies are a good choice to block rain and sun; natural areas in the woods also allow the animals to escape from the elements. Awnings or roof overhangs can provide a good place for the goats to get out of the weather yet stay out of confined areas. A tree-lined pasture will also provide respite from the sun. Be assured limbs will be trimmed as high as the animals can reach.

If you travel to other farms, be aware that you bring back new sources of disease on your boots, truck, and wagon tires. Biohazards have become something to be aware of and are a real threat to your farm. If you travel to another farm, wear rubber boots and wash them in a sanitizing solution before returning to your land. Disposable plastic boot covers are a good thing to keep on hand. Upon returning to your farm, wash your truck and equipment to avoid the introduction of new microorganisms. You cannot be too careful.

Vaccinations. As a human takes immunizations against certain predicated diseases, there are recommended vaccines for goats as well. Vaccines work to stimulate the production of specific antibodies to combat particular threats. The following vaccines are currently available: enterotoxaemia and tetanus (known as a CDT vaccine), pneumonia, foot rot, caseous lymphadenitis, and rabies.

While vaccination does not guarantee the animal will not contract these conditions, they definitely buy some insurance against major outbreaks. Some of these vaccines are not compatible with each other, so check with your veterinarian to set the protocol.

Wormers. Worms are part of life with goats. Unfortunately there is not one product available that will address all the parasites that select goats as hosts. Fecal samples can be examined at home, but you'll need a microscope. These samples are the only way to truly know what parasites are present. If you understand the nature of the beast, then you can treat for that specific parasite. However, many farmers choose a shotgun approach, rotating wormers and hoping to hit the targets. Work with your veterinarian or extension specialist to formulate a rotation program.

Internal parasites that are common to goats are *Haemonchus* (wire worm), *Ostertagia* (brown stomach worm), *Trichostrongylus* (nematode or round worm), and *Cooperia* (intestinal worm). Tapeworms and coccidia are also problematic. Dry conditions incite parasitic overloads, as do rainy, wet conditions. It is simply a given: if there are goats, there are worms and other pests to deal with. And if it isn't one thing, it's another. Goats are susceptible to lice, ticks, and other exterior parasites, too. Careful monitoring is all a part of goat husbandry.

Wormers are available in several forms. Some are in a pelleted feed supplement, some as a pour on, and others as a drench or as an additive to the water supply. Check labels and contact manufactures to see about the withhold time after worming. Usually there is a time when milk will need to be discarded and when meat should not be consumed, along with other precautions. Be sure to follow these instructions.

Hoof trimming. Unless your goats are on hard, rocky surfaces, hoof trimming will be a routine maintenance procedure. Without care, the hooves will become overgrown, turn into pointy elf shoes, or overlap and trap manure inside to form a pocket of infection. Before you start, purchase a good pair of trimmers and study kids' hooves and see how they are shaped.

When you're ready to go, catch the mature goat and either press him into a fence to get a good hold or tie him to the fence or the stanchion. I always stand straddle legged over the goat and bend down to pick up one hoof at a time. (If you are fortunate enough to have a turning table—good for you! If not be prepared to stand on your head—at least for a portion of the day.) Grasp one leg firmly, and bring it up for inspection. Using the trimmers, cut the overgrowth away. Don't make the cuts too deep or the hoof will bleed. It is better to do a little at a time, come back at a later time, and do a little more than to end up with a lame goat. Routine maintenance will avoid real problems.

If the animal has hoof rot you will know it and soon as you cut into the hoof. A rank odor will be present, often with evidence of infection. Copper sulfate (10%) is a good treatment. Buy the pellets, add water in a coffee can, dip the goat's hoof in, and let him go. However, if infection is present in a major way, then a round of antibiotics may be required to clear it up. Bad hooves can cause infection that travels up the leg and can result in lameness and eventually death. So take care of those hooves!

Medications. Make sure to store the medications in a locked cabinet or to refrigerate them according to directions. Keep them away from children. Animal preparations can be just as lethal as medications intended for humans. Don't use them lightly in animals, and do not use them for human consumption.

☙ Disease ☙

According to Dr. Charlotte Clifford-Rathert, a veterinarian and small ruminant specialist with the Lincoln University Cooperative Extension and Research, the following goat diseases are highly infective within a herd and require intense management for proper control; however, these diseases are not reportable. These diseases can be very costly to producers and also have a zoonotic (contagious to man) potential.

Sore mouth (contagious ecthyma, Orf)

This disease affects sheep and goats and is caused by a parapoxvirus. Lesions most commonly occur on the mouth and face but can also occur on the feet, teats, and genitalia. The poxvirus is present worldwide and can remain infective in the scabs in the environment for months to years.

The virus is spread by direct or indirect contact from environmental contaminants. The virus enters through abrasions (scrapes) or wounds of the mouth, teat, feet, or genitalia. It then localizes in the tissues and is shed in the scab. Animals that are kept in the same area are at the greatest risk. The infection is self-limiting, with most animals developing protective immunity. However, reinfection is possible.

Clinical signs. Early signs are small bumps or blisters on affected skin, usually around the mouth. Thick brown to black crusts form and are most evident. Lesions typically resolve in 14 to 21 days. Nursing lambs or kids are most likely to spread the disease to udders of the susceptible ewes or does. Oral lesions may become so severe that they cause an animal to stop eating.

Diagnosis. Observation of clinical sign and skin biopsy.

Treatment. Treatment of individually affected animals is not provided unless lesions are severe. Consult a veterinarian.

Prevention. Put control measures into practice immediately. Affected animals should be separated from all the other animals. Prevent the scabs from falling off into the environment. Vaccines are available but not recommended in disease-free herds because the vaccine contains live virus and poses a contamination threat.

Contagious ecthyma is highly zoonotic and may produce lesions on the hands or fingers of the person handling infected animals. Therefore it is extremely important to practice good hygiene.

Pinkeye

Pinkeye, also known as viral keratoconjuctivitis, has been reported in goats. It can be a sequel to bluetongue virus infection. Pinkeye in goats is a mycoplasmal disease. A surface infection with the mycoplasma organism in the eye can cause pinkeye in goats and sheep. The organism can also enter the blood stream and cause septicemia, abortion, respiratory problems, and arthritis in multiple joints. Flare-ups occur in times of stress, overcrowding, or kidding. Pinkeye can be spread by direct or indirect contact with infected animals or body fluids from infected animals. Newborn goats can spread the organisms from the mother's mouth to her udder and in turn become infected by ingesting contaminated milk.

Clinical signs. Cloudiness of the cornea. A mycoplasma infection should be suspected in goats with severe pneumonia.

Diagnosis. Mycoplasma strains can be identified by bacterial culture or staining of discharges from the eye. Consult a veterinarian for diagnosis and treatment.

Prevention. No vaccine is available in the United States. Providing good fly control, preventing stress and overcrowding, and separating infected animals from healthy ones will help prevent the spread of disease.

Johne's disease (paratuberculosis)

This chronic disease causes a wasting body condition, without diarrhea, in goats. It is caused by the bacteria *Mycobacterium avium* subspecies *paratuberculosis* (MAP) and is spread by the fecal-oral route. Young animals are more susceptible to the disease than adults. It is also transmitted through milk and placenta, therefore predisposing the young of infected animals to the disease, especially those showing symptoms. After an animal is exposed, it will either clear the organism or develop a chronic persistent infection, depending on its immune status.

The most consistent clinical sign is chronic weight loss that is not caused by parasite infection despite a good appetite. It eventually results in death. Clinical signs of this disease are so subtle that it may take months or years to realize there is a problem. Meanwhile an infected animal can be shedding the organism in its feces, contaminating the environment and other animals in the herd.

Diagnosis. Culture of the organism from feces is the official test; however, this process takes up to 8 weeks or longer. Newer diagnostic methods available are blood testing by the enzyme-linked immosorbent assay (ELISA) and agar-gel immunodiffision (AGID). Both these tests are rapid and results are available in 48 hours. These tests can be useful as screening tools for whole-herd management, with ELISA as a whole herd test and AGID as an individual test.

Treatment. There is no effective treatment, so prevention and control are very important. Preventing the introduction of Johne's into a herd can be very difficult. The best prevention is to maintain a closed herd, but this action may not be practical. Blood testing of animals is helpful when purchasing noninfected animals but is not 100% reliable. Ask about herd history before purchasing, and culture newly purchased animals every 6 months due to the subclinical nature of the disease.

Control. Control is a combination of management, education and screening periodically. Control of the fecal-oral route of infection is a vital part of any herd management plan. Consult a veterinarian or livestock extension agent to help develop effective management plans, and confirm/cull any suspect animals. A clean environment is important especially in the kidding barn and pastures. Using above ground feeders and waterers and cleaning the does' udders before nursing young are two ways that will help control the fecal-oral route of infection.

Caprine Arthritis-Encephalitis Virus (CAE)

CAE is a chronic multisystemic disease in goats. Infection is widespread, and arthritis in more than one joint is the most common clinical sign. Infection occurs by ingestion of fluids that contain infected cells from an infected animal to an uninfected animal. The most common means of transmission is the ingestion of colostrums by kids nursing infected does. CAE can also be spread by breeding, contaminated dehorning equipment, and at parturition.

The target tissues of the CAE virus are joints, mammary glands, lungs, and brain. The disease results from inflammation induced by the reaction of the immune system to the virus. Goats can develop a blood titer in 2 to 8 weeks but may not show clinical signs for years.

Clinical signs. Infected goats show progressive arthritis in animals over 6 months of age, usually noted in the front pastern joints, with chronic progression over the years.

Diagnosis. Routine diagnosis is based on specific serological (blood) testing called agar gel immundoffiusion test (AGID) or polymerase chain reaction assay (PSR).

Treatment. There is no treatment; affected animals are a source of infection to others. It is recommended to cull infected and positive animals in order to eradicate the disease on the farm; otherwise, rigorous management is required.

Caseous Lymphadenitis (CLA)

This can be a devastating disease caused by the bacteria *Corynebacterium pseudotuberculosis*. It is more common in sheep than in goats and causes abscesses of the skin and subcutaneous lymph nodes that will break open to the skin's surface and contaminate the environment. This disease may affect the animal internally, most commonly the respiratory system, causing long-term respiratory problems. It can also spread to the abdominal lymph nodes (weight loss), central nervous system (neurological signs), and mammary gland (mastitis).

The organism can survive for prolonged periods of time in dark, damp areas, soil, and manure. The most common means of infection is by injury with contaminated shears, dip tanks, needles, and puncture wounds. Abscesses form then break and drain into the environment (feed bunk or water buckets or on the ground), predisposing the next animal to come along to become infected through a wound.

Clinical signs. Signs of infection are superficial lymph node swellings with draining tracts. An onion ring appearance shows on the abscess if surgically removed; a pale greenish, cream-colored, pasty discharge drains from the access when ruptured.

Diagnosis. Diagnosis is based on serological (blood) testing and culture of organism from abscess.

Treatment. The abscesses should not be opened within the vicinity of the other animals. Isolate affected animals for treatment. If possible, have the abscesses surgically removed to reduce contamination of the environment. If the abscesses have already ruptured, the animals should be isolated, and the abscess should be flushed with antiseptic solution (3% iodine or 2% chlorhexidine) and packed with antiseptic saturated gauze. In the event of rupture, it is best to seek the help of your veterinarian.

Remember, where you buy your animals will often predict the amount of problems you will have. Many sick animals have been inherited from auctions or disreputable animal traders. Deal with people you know or those who are recommended by various goat associations, local veterinarians or friends. Bringing in an animal with a health condition can threaten the whole herd.

The Art of Cooking Goat Meat

Goat meat is gaining popularity in the United States with production of meat goats increasing annually. According to the U.S. Agricultural Census data, in 2002, 74,908 farms produced 1,938,924 head. By the next census done in 2007, the numbers increased to 123,278 farms producing 2,601,669 head (U.S. Department of Agriculture 2007 Agriculture Census Data). This demand for goat meat is a result of the changing population in the United States. Goat is the meat of choice for various festivals and religious celebrations. As people become acquainted with the flavor and versatility of the meat, the word is spreading and markets are continuing to increase.

When selecting an animal for food preparation, the size and age of the goat will dictate the intended use and recipe. Young goat can be cooked rather quickly; a goat 6 to 9 months old is ideal for roasting. Older animals are ideal for stewing, slow cooking, or grinding into burger. Goat roasts are popular for celebrations. Typically a goat roast will require a spit and outdoor fire pit. Goat chops, roasts, burgers, and sausage are also delicious. If I had to compare goat to any other meat, I'd say it has a lot of the same qualities of venison, and their texture and cooking methods are closely related.

I will not go into the butchering of the animal for the purposes of this book. Check online resources if you want to learn to butcher or purchase an animal live through a meat processing facility. For those who decide to produce meat goats on the farm, your local extension agent can supply the facts and regulations for processing in your state.

As common sense prevails, the tenderest animal will be youngest. Tough meat comes with age, so consider this when purchasing goat meat. As with beef cattle, ages and weights correlate with good steaks as opposed to the senior citizen stock that is turned into hamburger. If a male goat has not been castrated, there will be a musky flavor associated with the meat. Some consumers like this flavor; others do not care for it. Again, question the producer or supplier to learn what type of meat and the age of animal you are buying if possible. Roasted goat meat is lower in fat than beef or pork and higher in protein.

Cook ground goat meat until it reaches a temperature of 160° F. For roasts and steaks, 170° F is considered to be well done.

When cooking goat—other than on an outdoor roast on a spit—I always like to cook the meat with onions, garlic, or peppers to offset any gamey flavor. Therefore, most recipes will include those ingredients. Goat meat goes particularly well with Tex-Mex or Latin American favorites. Fajitas, burritos, tacos, and similar foods are complemented by the use of goat. Again, the traditional ingredients of pepper, onions, and garlic serve to enhance the flavor of the meat. Another way to go is to add ingredients on the sweet side such as apples, molasses, honey, or a sweet BBQ sauce. When cooked correctly, most people will have no idea they are eating goat meat.

Let's start with the youngest: cabrito. Made popular in Latin America, this dish is created from a young, milk-fed goat, 1 to 3 months of age.

Spit-Roasted Goat

Ingredients
2 kids, up to 10 pounds each, dressed and ready to cook
2 bottles Mojo Criollo marinade (Goya has a great product.)
Serves 20–25

1. Place the meat in a pot, and pour the Mojo Criollo over it. Marinate the meat for 12 to 24 hours, turning every 4 to 6 hours to evenly coat the meat.

2. Prepare a cooking fire outside with a split to roast the meat. Mesquite chips are great for this step, but apple, cherry, or other sweet woods are good for roasting, too. Allow the fire to burn down to coals. Secure the meat on the spit, or place it on a cooking rack over the fire. Turn the meat frequently to prevent burning and promote even cooking.

3. Cook the meat for about 3 hours or until the internal temperature of the meat (measure at the hip joint, the thickest point) reaches 170° F for well done meat. Baste with Mojo Criollo throughout the cooking process.

4. Remove the meat from the spit or cooking rack and carve. Serve with black beans, rice, Pico de Gallo (recipe follows on page 92), and tortillas.

Pico de Gallo combines the best of summer's flavors. Fresh cilantro, tomatoes, and onions blend together in perfect harmony. This bright and colorful relish goes well with eggs, meat, and of course, cheese. This is a Hurst family recipe.

Pico de Gallo

Ingredients

1 bunch green onions
1 small white onion
1 green pepper, seeded
6 ripe plum tomatoes
1 can Ro*tel tomatoes with green chilies
1 bunch cilantro
2 tablespoons lime juice
½ to 1 teaspoon salt
Yields 4 cups

1. Using a food processor, place the green onions, white onion, and green pepper in the bowl. Process until chopped.

2. Add whole plum tomatoes, Ro*tel, and cilantro. Process until blended.

3. Stir in lime juice and salt.

Fried Plantain

Ingredients
2 ripe plantains
½ teaspoon powdered ginger
½ teaspoon cayenne pepper
Cooking oil
Makes 4 servings

1. Purchase two plantains and wait until the peel turns black. Peel and cut into ½-inch slices.
2. Heat oil in a sauté pan, and fry the plantain until brown.
3. Remove the plantain from the heat, then drain and sprinkle with ground ginger and cayenne pepper.

Serve with a dollop of sour cream and as a side dish to spicy food.

Jerked Goat

This recipe is best when grilled.

Ingredients
2 goat necks or 2 goat hind legs

Jerk Sauce
1 tablespoon ground allspice
1 teaspoon cayenne pepper
1 teaspoon black pepper
1 teaspoon pumpkin pie spice
4 cloves garlic, minced
1 to 3 habanero peppers, seeded and chopped (wear gloves)
1 cup finely chopped onion
½ cup brown sugar
¼ cup apple cider vinegar
Juice of 2 limes
Serves 12

1. Combine spices. Add minced garlic, chopped pepper, and onions. Stir in brown sugar, vinegar, and lime juice. Make a paste.
2. Using a sharp knife, make holes in the goat meat (this is the jerking). Wear gloves to rub the paste over the meat. Place the meat in plastic bags, and marinate overnight.
3. Grill the meat for approximately 1 hour over a low flame. The juices should run clear when the meat is cooked through. Add salt to taste after grilling. Chop the meat into pieces and serve with Jamaican bread, grilled pineapple, and Fried Plantain (recipe above).

Goat with Apple

Ingredients
One 3 to 4 pound goat roast (very lean)
1 teaspoon salt
1 teaspoon ground ginger
10 peppercorns
10 whole cloves
1 cup red wine
1 cup apple cider vinegar
¼ cup brown sugar
2 onions, sliced thin
1 bay leaf
2 tablespoons vegetable oil
¾ cup crushed gingersnaps
Serves 4–6

1. Rub the roast with salt and ground ginger. Place the remaining ingredients in a saucepan, and cook over high heat to a boil. Cool.

2. Place the roast in a large bowl, then pour the marinade over the roast. Cover the roast with plastic wrap, and refrigerate for 2 to 3 days. After that time, remove the roast and reserve the marinade.

3. Pat the roast dry, and brown it in a Dutch oven with 2 tablespoons of vegetable oil. Brown thoroughly.

4. Pour the marinade over the roast, cover, and simmer 3 hours or until the meat is tender.

5. Remove the meat and strain the marinade, discarding the onions and other solids. Stir in ¾ cup crushed gingersnaps, and simmer until the gravy is thickened.

Slice the roast and serve it with the gravy, Cooked Apples (recipe follows), and herbed new potatoes.

Cooked Apples

Ingredients
6 large Gala apples
2 tablespoons butter
1 teaspoon apple pie spice
2 tablespoons sugar (brown or white)
Serves 4

1. Slice and core the apples, leaving the skin on. Heat large skillet and add butter.

2. Stir in the apples, coating them with the butter. Cook until tender (15 to 20 minutes over medium heat). Turn apples throughout the cooking process to prevent overcooking.

3. Add 2 tablespoons of sugar just before serving.

Jamaican Bread

Ingredients
2 tablespoons dry
  yeast
1 teaspoon sugar
¼ cup warm water
1 teaspoon salt
¾ cup warm milk
3 cups flour
1 egg, beaten
1 stick of butter melted
  and divided into two
  portions

Caribbean flavors combine to create a jerk sauce. Hot peppers, brown sugar, vinegar, and lime juice tenderize and flavor the meat. Jamaican Bread and pineapple cool the heat from the sauce, and Fried Plantain with sour cream adds the finishing touch.

1. Dissolve yeast in warm water, stir in sugar, and let sit for 10 minutes until bubbly.

2. Add salt, warm milk, egg, and ¼ cup butter to yeast mixture.

3. Add flour 1 cup at a time, working it in to form the dough.

4. Turn the dough out of the bowl onto a floured work surface and knead until smooth and elastic. Add more flour if it is sticky.

5. Place the dough in large, oiled bowl and cover with a damp kitchen towel. Let raise for about 50 minutes. Make sure the room is warm so the dough will rise properly (it should double in size).

6. Remove the dough from the bowl and divide in half. From each half form five equal pieces.

7. Using a rolling pin, roll the pieces into 6-inch-diameter circles.

8. Brush dough with melted butter and fold in half, making half moons.

9. Place the bread on a parchment-lined baking sheet and let rise again for 30 minutes.

10. Bake at 425° F for 15 minutes.

Chevon in Mustard Sauce (Chevon à la moutarde)

Ingredients

4 goat chops, cut 1½ to 2 inches thick
4 tablespoons butter
1 sweet onion, chopped
½ cup Marsala wine
½ cup chicken broth
½ cup Dijon mustard
½ teaspoon thyme (Lemon thyme is recommended.)
½ cup crème fraiche
Chopped parsley
Serves 4

1. Brown the chops in butter in a sauté pan. Use a low heat to avoid scorching.

2. Remove the chops from the pan, and stir in the chopped sweet onion. Cook the onions until tender; set aside.

3. Pour Marsala and chicken broth into the sauté pan, and then stir in mustard and thyme. Bring the mixture to a boil.

4. Add the chops, and spoon the sauce over them. Cover, and cook over a low heat for 30 to 40 minutes.

5. When the meat is cooked through and very tender, remove the chops from the pan.

6. Continue to cook the sauce until only half remains. Add the crème fraiche and parsley, heat through. Spoon the sauce over the goat chops.

Serve with smashed potatoes and sweet green peas.

Mustard sauce is a classic French sauce often used with rabbit. It is also a tasty complement to goat chops!

Kabobs are a summertime grilling classic with a Mediterranean flair.

Goat Kabobs

Ingredients

2 to 3 pounds tender cuts of goat meat, such as the tenderloin

Marinade, such as soy sauce, apple juice, or Mojo Criollo

4 sweet onions

Green peppers

Whole mushroom caps

Cherry tomatoes

Garlic salt

Serves 4

1. Cut the goat meat into 2-inch cubes.

2. Slice the onions into wedges; seed, core, and slice the green peppers. Wash the tomatoes and mushrooms, and remove the stems from the mushrooms.

3. Marinate the meat in soy sauce, apple juice, or Moho Criollo for 12 hours. Remove the meat from the marinade, and pat dry.

4. Thread the meat and vegetables on skewers; sprinkle with garlic salt to taste. Cook over medium flame on a BBQ grill.

   Serve with Grilled Sweet Potatoes (see recipe, page 99).

Meat Loaf with Ground Goat Meat

Ingredients
1 pound ground goat meat
2 eggs
½ cup Pico de Gallo
1 small onion, chopped
2 cups crushed saltine crackers
Chipotle Ketchup (recipe follows)
Serves 6

1. Combine the ingredients in the order listed, up to the ketchup.

2. Form the mix into a loaf, and place it in a baking pan. Bake for 45 minutes at 350° F.

3. Add ¼ cup Chipotle Ketchup (purchased or homemade, recipe below) on the top of the meat loaf, and return it to the oven for 5 minutes. Remove it from the heat, and let it stand for 5 minutes.

   Slice and serve.

Chipotle Ketchup

To 1 cup of your favorite ketchup, add 3 tablespoons canned chipotle in adobo puree. Add a dash of cinnamon, salt, and pepper and 1 tablespoon honey, and stir it into the ketchup adobo mix. Refrigerate. Yields a little over 1 cup.

Here's a good ol' American staple with a twist: goat meat instead of the usual pork and beef.

Grilled Sweet Potatoes

Ingredients
2 to 3 sweet potatoes, sliced ½-inch thick
Olive oil
Cinnamon
Salt and pepper
Serves 4

1. Brush the sweet potatoes with olive oil, and cook them on the top rack of the BBQ grill. Turn gently, as necessary to keep from scorching.
2. Cook for 25 to 30 minutes; add cinnamon, salt, and pepper just before serving.

Curried Goat

Ingredients
2 pounds goat meat, cubed
2 tablespoons curry powder
1 clove garlic, crushed
2 onions, chopped
2 tomatoes, chopped
½ Scotch bonnet (habanero) pepper, chopped
2 tablespoons butter
Makes 8 servings

1. Combine curry, garlic, onions, tomatoes, pepper, salt, and black pepper.
2. Place meat cubes in mixture, and cover with plastic wrap. Allow to marinate for 1 hour; overnight is better.
3. Remove the meat from the seasonings. Brown the meat in butter.
4. Add the vegetables, and sauté until tender.
5. Add 2 cups of water, and cook the vegetables until they are tender and the sauce is reduced to a thick stew.

Serve with hot cooked rice, chutney, raisins, nuts, mango, and Fried Plantain.

Curried Ribs

Goat ribs can be tough, depending upon the age of the goat. For an older animal, parboiling is recommended. Simply add the uncooked ribs to boiling water with a splash of vinegar or beer, boil for 15 minutes, then proceed with the recipe.

Ingredients
2 tablespoons vegetable oil
4 to 5 pounds goat ribs
1 tablespoon Worcestershire sauce
1 teaspoon salt
½ teaspoon cayenne pepper
8 ounces beer
¼ cup brown sugar
¼ cup tomato paste
Hot pepper sauce, to taste
Makes 6 to 8 servings

1. If ribs are parboiled, pat them dry before placing them in hot oil.

2. Add oil to a large skillet, and brown the ribs. Remove the ribs from the skillet, and sauté the onions until tender.

3. Place the ribs back in the skillet with the onions; add additional ingredients and pour over ribs.

4. Cover and cook at 350° until tender, about 1½ hours.

Serve with hot sweet potato fries.

In a new twist on an old favorite, goat meat is used as the base in this tasty stuffed pepper. A bit of mozzarella adds another new element. In my opinion, everything is better with a little bit of cheese!

Stuffed Peppers

Ingredients

4 multicolor bell peppers
2 tablespoons olive oil
1 pound goat burger
2 cloves garlic, minced
1 sweet onion, chopped
Salt and pepper
1 can tomato sauce
1 teaspoon sugar
¼ teaspoon dried basil
½ teaspoon dried oregano
1 cup cooked rice
Fresh chopped basil
Cherry tomatoes
Fresh balls of mini
    mozzarella cheese
Makes 4 servings

1. Cut the tops off the peppers and seed them, being careful to leave the peppers intact.

2. Sauté the burger, garlic, and onions in olive oil. Remove from heat.

3. Stir in cooked rice, tomatoes, basil, and oregano, and tomato sauce.

4. Spoon the mixture into the peppers, and arrange the peppers in a baking dish.

5. Bake for 45 minutes at 350° F; cover the dish with foil or a lid to keep the steam in the dish.

To serve, place the peppers on a serving dish. Garnish with sliced cherry tomatoes, basil, and mini mozzarella balls. Serve with garlic bread.

This hearty goat stew is ideal for those chilly autumn days.

Goat Stew with Chili Sauce

Ingredients

2 pounds goat meat, cubes or left-over pieces from butchering
4 guajillo chilies (red dried Mexican chilies)
3 poblano chilies, seeded and chopped
½ cup hot water
3 garlic cloves, minced
1 teaspoon ground cinnamon
½ teaspoon ground cloves
2 tablespoons vegetable oil
Salt
Potatoes, sliced ½-inch thick with peel left on
Makes 8 servings

1. Tear up the dried guajillo chilies, and soak them in ¼ cup hot water for 30 minutes.

2. Add garlic and spices, and process in a food processor or blender until smooth.

3. Using a large skillet, add the oil and cook the meat until it is evenly browned.

4. Layer the sliced potatoes and poblanos over the meat, and cook for 15 minutes, covered.

5. Pour the chili mixture over the top; add salt to taste. Continue to cook, 45 to 50 minutes more over medium heat, until the potatoes are tender.

Serve with rice and corn bread.

Gyro with Goat Meat

Ingredients
Tender cut of goat meat, sliced very thin (2 pounds)
¾ cup balsamic vinaigrette salad dressing
3 tablespoons lemon juice
1 tablespoon dried oregano
½ teaspoon freshly ground black pepper
Makes 4 to 6 servings

1. Combine salad dressing, lemon juice, oregano, and pepper.
2. Place the goat meat in a plastic bag, and pour the salad dressing mixture over the meat. Refrigerate for at least an hour.
3. Remove the meat from the marinade, and place it in a large skillet. Cook the meat quickly; add 4 tablespoons of the marinade to the skillet just before removing the meat. Serve with yogurt sauce, lettuce, sliced onions, sliced cucumbers, and black olives on pita bread. Top with feta cheese.

Yogurt Sauce

Ingredients
1 cup plain yogurt
½ cup mint leaves, chopped
1 garlic clove, minced
1 tablespoon fresh lemon juice
Salt and pepper to taste
Yields 1 cup

Stir the ingredients together, cover, and refrigerate for 1 hour.

Ingridsl/Shutterstock.com

Rosemary Raisin Pecan Crisps

This delicious snack comes from my friend, Mary Kay Robinson, a fellow cheese affectionado.

Ingredients
2 cups flour
2 teaspoons baking powder
1 teaspoon salt
2 cups buttermilk
¼ cup brown sugar
¼ cup honey
1 cup raisins
½ cup chopped pecans
¼ cup sesame seeds
¼ cup flax seed, ground
1 tablespoon chopped fresh rosemary

1. Preheat the oven to 350° F.

2. In a large bowl, stir together the flour, baking soda, and salt.

3. Add the buttermilk, brown sugar, and honey; stir lightly.

4. Add the raisins, pecans, sesame seeds, flax seed, and rosemary. Stir just until blended.

5. Pour the batter into one 7 x 10-inch baking pan that has been coated with nonstick spray.

6. Bake for about 45 minutes, until golden brown and springy to the touch.

7. Remove the bread from the pans, and cool on a wire rack.

Once cool, cut the bread in 10 2⅓-inch pieces. Wrap each piece individually in plastic wrap then wrap in foil. Freeze. Can be frozen for 1 month.

To Make the Crackers. Keep the bread frozen so it is easy to slice. Slice the loaves as thin as possible, and place the slices in a single layer on an ungreased cookie sheet. Preheat the oven to 200° F, and bake the slices for about 15 minutes; then flip them over and bake another 10 minutes, until crisp and deep golden. Do not place the crackers in a bag or other airtight container until they are completely cooled. These crisps do not store well when baked a second time. So make, bake, and enjoy.

Pecan-Crusted Goat Chops

Ingredients
6 1-inch-thick goat chops
1 cup finely chopped pecans
½ cup crushed crackers
1 teaspoon garlic salt
¼ teaspoon freshly ground black pepper
½ cup all-purpose flour
2 eggs
½ cup milk
¼ cup vegetable oil
Makes 6 servings

1. Combine pecans, crackers, garlic salt, and pepper.

2. Blend milk and eggs.

3. Dip the goat chops in the milk and eggs, then press them into the pecan and cracker mix. Coat thoroughly.

4. Pour the oil in a large skillet, and place it over medium flame. When the oil is hot, gently place the chops in the skillet. Let them brown thoroughly on one side, then turn the chops and cook the other side.

5. Continue to cook for about 25 minutes.

A flavorful goat chop with a pecan crust. The applesauce is a favorite recipe. Simply peel Jonathan's or other cooking apples, add a small amount of water, and cook until tender. Mash with a potato masher and add sugar to taste. The secret ingredient—red hots! Add by the handful for a cinnamon-y zing and great color.

Black Bean Chili

Ingredients
2 tablespoons vegetable oil
1 pound ground goat meat
1 onion, chopped
2 cloves garlic, minced
2 (15 ounce) cans black beans, undrained
1 (14.5 ounce) can crushed tomatoes
1½ tablespoons chili powder
1 tablespoon dried oregano
1 tablespoon dried basil leaves
1 tablespoon red wine vinegar
Makes 8 servings

1. Sauté onion and garlic until tender in a Dutch oven.

2. Add ground goat meat, and cook until well browned.

3. Add black beans, crushed tomatoes, spices, and red wine vinegar; stir well.

4. Cook over medium to low heat for about 1 hour.

Serve with corn bread. Garnish with sour cream and cheddar cheese, and add pepper jelly
   on the side.

Black Bean Chili with cornbread is sure to warm the heart and soul on a mid-winter's day. Goat meat is complemented by many spices and black beans. Cornbread makes everything better and is a regular staple on the Hurst menu.

Goats for Fiber

*F*iber animals have to be some of the most beautiful beasts to inhabit the earth. Sheep, llamas, alpacas, angora rabbits, angora goats, and cashmere goats give us the most luxurious fabric known to man. A mature angora goat will produce 10 to 15 pounds of mohair each year, making it an ideal addition to a fiber-producing herd.

⚘ Mohair ⚘

Angora goats give us mohair. They are known for their curly, silken locks and angelic appearance. This breed is typically docile, most concerned with browsing and producing hair.

The fiber from angora goats has a shimmery, silken texture known for its durability. Military uniforms of the past, high-end upholstery, and even covering for airplane seats have been made of fabric containing mohair. Spinners and weavers value this fiber, making the kid curls into a spectacular yarn. When spun with a trained hand, the tiny curls create slubs and extensions in the yarn that then are highlighted in the crafting of a garment. The tiny curls remain visible in the yarn, producing a texture that is unsurpassed. Manmade fibers cannot duplicate the delicacy of this finely spun yarn. Conversely, the yarn can also be spun into thick, ropelike material.

Highly durable, this type of yarn can be used for projects such as purses or bags that will receive heavy use. Doll wigs and Santa beards are often made of mohair, too. If you are lucky enough to find an antique Santa suit, chances are the beard will be made of luxurious mohair locks. Often the trim on the suit will be made of fabric containing mohair.

Mohair at its finest—as the perfect beard and hair for a wizard. This magician began as a cloth body and resin face. Choosing fabrics and accessories can turn the basic form into a jolly old Santa or other personality-packed beings. To apply the hair, begin at the back of the head at the neckline. Simply work up to the middle of the head, adding one lock at a time. Then, with the figure facing forward, layer from the middle of the head to the top of the forehead. For the beard, work from the chin to the lower lip. This creates a realistic look and appears as natural hair growth.

Mohair locks. White is most common and will range from true white to ivory. The silvery grey is called "blue" in the language of goats.

With all these curls, angoras have a tendency to pick up debris such as cockleburs, small sticks, and other materials in their locks. Some farmers who raise fiber goats will put a lightweight coat over them to keep their hair clean. Maintenance of their browsing area can prevent some debris collection, but it is largely inevitable.

Shearing

Angora goats are sheared twice a year—once in early spring and once in the fall. Watching the weather is important, because without their coats, the goats will chill if a cold snap happens unexpectedly. I keep a stack of children's sweatshirts with the sleeves cut off in the event a kid becomes chilled. Of course my neighbors talk about that crazy goat lady with her goats in sweatshirts; however, this effort has proven to be a viable means to warm up a chilled goat and reduce the risk of loss. Commercially made goat coats are also available.

Shearing is an acquired skill, and it is best to go to a shearing school or to hire a skilled shearer to take care of this task. Without proper training and equipment, unskilled shearers can easily harm the animal with the blade of the shears. Mohair does not contain lanolin, as wool does, so the goats are more difficult to shear. It takes time and practice to learn this skill and perform it with certainty.

One of the worst things that can happen during shearing is to shear off a nipple. This unfortunate mishap causes problems down the line when the goat becomes pregnant again. There is no way to relieve the pressure of the milk, which will still be produced and present in the udder. Most good shearers will cover the teats with their fingers while working the belly to avoid this disaster.

General cuts from the sharp blades are also a risk, and it is not uncommon to have to stitch up a wound. Blood stop powder is available at farm supply stores and should be kept on hand for shearing day and other accidents. This substance coagulates the blood quickly and helps to stop any bleeding.

Collecting Mohair

Shearing twice a year allows the angora coat to grow to a (staple) length suitable for carding. Carding may be done manually using wool carders or sent to a carding mill. First the fiber should be manually picked through to remove any debris or vegetation. It is important to keep the curls as intact as possible. Kids will have the most highly prized and valuable fibers.

Currently mohair is priced at $3 to $12 a pound. The great variance is attributed to the length of the staple; whether the mohair had been skirted (vegetation removed), cleaned, or washed; and whether the fiber is from a kid or a mature animal. Kid hair brings the highest price. Naturally colored mohair is priced at $10 to $25 a pound with the same variances applied.

Older goats lose the tightness of the curl, and their fiber does not have the same sheen or texture. It will develop "kemp," which is a far less desirable fiber. Kemp is somewhat matted,

Artists Kim Carr and Craig Yenke

Goats are the favorite subject of many artists. Those who understand goats capture their personalities on canvas, in sculpture, and on film. Two such artists are Kim Carr and Craig Yenke (www.craigyenke.com). Both have the ability to connect with the animals they use in their art, capturing the essence of each creature. Kim works as photographer in addition to her fulltime work as a farmer. She describes herself as a farmer/photographer, and each vocation complements the other. At the heart of Kim's work is an innate understanding of the animal. Whether it is dogs, llamas, her favorite hen, or goats, Kim finds the soul of the animal and brings it directly into her photos. It is a rare gift. Kim's photos are available as framed prints, note cards and Christmas cards. (www.kimcarrphotography.com).

Craig Yenke creates lifelike sculptures of almost every breed of animal you can imagine. From horses, to moose, turkeys and of course, goats, Craig's work is incredibly lifelike. Inspired by the past, Craig's art evokes a bygone era when the quality of craftsmanship was the most important thing. Craig stresses the importance of collectors knowing his work is "fashioned 100 percent with my two hands—start to finish." Craig doesn't use any molds or premade pieces. "Handmade is the most important aspect to me," he notes. "I want every piece to be completely original and unique, displaying its very own character and realism."

Craig's love of animals becomes evident as one realizes that it takes up to 20 individual steps to complete just one piece. They are made from paper mache, cotton batting, and other uniquely selected material for the coverings. The results of his work have been referred to as the heirlooms of tomorrow. Craig has received critical acclaim for his work, and one of my favorite things is one of Craig's sculptures, a goat who sits on my desk each day.

Craig Yenke's work captures the heart of the animal in sculpture. This little fellow normally sits on my desk and oversees my daily activities. He is so lifelike it is as if he could eat the excess paperwork.

and while it can be carded and spun, it does not have the attributes associated with mohair.

Another term used regarding the fiber is "stain." Mohair from the goat's rear quarters often has a dark stain. Many spinners capitalize on this natural coloring and do not dye this hair but leave it as is. After washing, a reddish color will remain, and the color variation is unique.

To clean mohair, first pick out the debris. Separate the stain from the rest of the fiber, and set it aside for special projects requiring colored hair. After picking out the debris, wash the hair. This procedure must be done with care. If the hair is agitated during washing it can felt, which results in a matted clump. Place the locks in a nylon bag (a laundry bag or lingerie bag), then soak them in a lidded container of hot water (145° F). Add a bit of dishwashing detergent or fiber wash to the water. Lay the bag in the water, pushing it in to submerge it. Let it soak for a few minutes, then turn the bag over and let it soak again.

After 45 minutes has passed, take a lock out of the bag to see if it feels clean or if it is still waxy. If it is waxy, remove the bag from the container, drain the water, and then add more hot water and detergent. Repeat the soaking until the fiber feels clean. When you are satisfied with the results, keep all the fiber in the bag and put it into your washing machine on the spin cycle. Do not add water; you are simply spinning out the excess water in the bag. When this step is finished, remove the mohair from the bag and place it on a screen to dry. When it is completely dry, it is ready to be carded. Some spinners will tease the locks a bit, which will add more texture to the yarn. To store fiber that has been washed but not yet carded, cloth bags are best. Suspend the bags, and keep them in a dry place where insects or mice will not have access.

Carding arranges all the fiber so that it all goes in the same direction. Fibers are worked on carders that resemble dog brushes. The metal teeth of the carders bite into the mohair (wool, cashmere, or other fiber) and align them for spinning.

To card using hand carders, arrange the locks on the carders and simply work one against the other, pulling the locks so the fibers all go in the same direction. To preserve the curl, pull lightly (some spinners work directly from the lock and avoid the carding process). Peel the prepared hair off the carders, and roll it in to a tubelike shape called a "rolag." Store the rolags in a basket, and set them aside for spinning.

A drum carder will automate the carding process. Some models have a hand crank and are not electrified. Others have a small electric motor that takes much of the work out of this step. Carding mills will process the fibers for a fee and send them back nice and clean, ready to spin. Fibers are returned as roving ready to be spun, premade yarns, or batts for felting.

Mohair takes dye exceptionally well. The colors remain vibrant and intense. The fiber may be dyed before or after spinning. Accents of the fiber will add color and texture to weavings, knitted, or crocheted items. Beautiful naturally colored fibers are available as well. Selective breeding has brought about a fabulous array from reds to blacks ranging from silver to rich grays.

⚜ Cashmere ⚜

Cashmere often holds the claim to the most luxurious fiber known to man. Amazingly, few actually associate the origin of this luxury

Wool carders are a bit of an investment but will last for years. The carders arrange all the fibers in the same direction. Whether you are spinning or creating batts for felting, carding is a necessary step. Antique carders show up frequently at estate sales. These are fine for display, but the wires may be old and rusted, so purchase new carders for best results.

Wordly Wares Fair Trade Products

The artists behind this purse and other projects are a wonderful group of ladies living in Lesotho, Africa. Elelloang Basali is a cooperative that was formed in 1997 in attempt to fight poverty and keep the traditional art of weaving alive. The five women are all from Lesotho. They spin, dye, and weave wool and mohair on the premises. They specialize in board loom tapestries and fine weaving.

There are numerous weavers throughout the country, most (if not all) of whom spin and dye the mohair themselves before weaving it into purses, scarves, hats, tapestries, and an assortment of other wonderful products. This particular group of women built their workshop and store out of tin cans. Their work caught the attention of Linda Fisher and Valerie Woods who import Fair Trade products and offer the items for sale through their company Worldly Wares (www.wordlywares.com).

Mohair is often the fiber of choice in the work of artisans from the Cooperative of Elelloang Basali of Lesotho, Africa. The integrity of the fiber is preserved in this weaving, with the overspun yarn creating in tight curls, resembling those of a kid goat.

Linda states, "Our visit with them was something to remember for a lifetime. They were so happy to see us and sang songs of praise to God for the small purchase of handbags and clutches that we made." Small, women-owned cooperatives are becoming a viable source of income for women all over the world. Not only does the much-needed income improve their overall quality of life, but the co-ops serve as a means of preserving the traditional arts and skills of various regions.

Lesotho is located entirely within the boundaries of South Africa. It is a nation formed and inhabited by a very peaceful population that originally sought to escape violence in South Africa. They did so by retreating into the mountains, at an altitude of over 4,000 feet above sea level. Naturally this is the perfect environment for surefooted goats, which became popular as livestock for the communities. Unfortunately the country has been devastated by the AIDS crisis, which has wiped out a large portion of the working population. In order to support their children and grandchildren, mothers and grandmothers formed cooperatives of weavers.

The women have captured the art of spinning the mohair, preserving the appearance of the curls of the fiber. The mohair is dyed with muted colors, allowing the fiber to be the primary focus of the work. Generations of skill, passed mother to daughter, make for a true piece of usable art.

These lovely swatches once lived life as a cashmere sweater. These are a sample of the yarns from Ellie's Reclaimed Cashmere. What patience to unravel sweaters for new creations—the ultimate recycling project.

with a goat. Those who consider goats to be revolting creatures will proudly sport a cashmere sweater or scarf, remaining oblivious to the producer of the silken fibers! It is said the best quality fiber is that which the goat actually sheds. This fiber is gathered by hand. Combing is another option for the small herd. However, most commercial operations today shear their animals annually rather than go on a scavenger hunt for the fibers left in the pasture. This fiber is treasured for its texture and sheen. With an adult producing about 4 ounces per year, it is small wonder that a cashmere garment is such an investment.

Cashmere wool fiber is obtained from Cashmere and other goat breeds. The goats produce a double fleece that consists of a fine, soft undercoat, or underdown, of hair mingled with a straighter and much coarser outer coating of hair called "guard hair." For the fine underdown to be sold and processed further, it must be de-haired. De-hairing is a mechanical process that separates the coarse hairs from the fine hairs. After de-hairing, the resulting cashmere is ready to be dyed and converted into yarn, fabrics, and garments.

An interesting product is now available: recycled cashmere. In this process, old cashmere garments are shredded and often combined with another fiber, then the fibers are respun to create beautiful yarns. Some fiber enthusiasts seek out cashmere garments and deconstruct them, saving the beautiful yarn to create one-of-a-kind skeins. One such entrepreneur is Ellen (Ellie) Joseph, a lifelong knitter who started her business 8 years ago after becoming a bit obsessed with the idea of knitting with cashmere. Having read an article about how generations past always reused yarn and finding the cashmere options in yarn shops limited, she began unraveling sweaters—first her own and then those she

Remembering First Grade

My first weaving experience goes back to first grade when the project was to make a simple weaving with construction paper. I still remember going over and under and thinking how much fun that was. My first attempt as an adult at a "crafty" project was basket weaving. I loved the way a flat piece of material could assume so many forms. When I went back to school in my 30s, I took a weaving class for fun. The second week of class, my instructor invited the class to her farm to see her angora goats. It was love at first sight! I already had dairy goats on my property, so two weeks later, I brought home my first two angora kids. It has been a love affair since that day. Angoras have the nicest personalities, are very docile, and the fiber is incredible.

found in thrift stores. She had such fun with the whole process that she decided to turn it into a business. Ellie's Reclaimed Cashmere was born.

Ellie goes regularly to the thrift stores in her area, on the hunt for the perfect color to complete a batch or that vintage golf sweater that someone finally parted with because of a few holes. Some days she hits the jackpot, and some days she goes home with an empty bag. She never knows what she'll find. Which is part of the fun! All the sweaters she uses are 100% cashmere and chosen because of color and the quality of the yarn. She doesn't dye any of the strands, preferring to create unique colorways with what she finds.

Once found, sweaters are washed in eco-friendly soap and then unraveled. Ellie takes the sweaters apart in pieces: front, back, sleeves, and collars. Much like handknit sweaters, they unravel top to bottom. Because they are knit with very fine yarn, most sweaters have quite a bit of yardage in them. Ellie produces about 150 pounds, the equivalent of 1,700 skeins, or about 250,000 yards a year, and she has unraveled more than 2,100 sweaters since she began her business!

Using three, four, or five different colored yarns, Ellie knits up miniswatches until she finds a color combination she likes. She then plies all the yarn from those sweaters together. Each colorway, or batch, is its own unique dye lot as she will likely never find the same sweaters again. The skeins are then washed again, tagged, and sent on their way to be reborn into a new knitting or crochet project. With each batch averaging over 2,000 yards, there is plenty for even the largest projects. Weight varies depending on the number of strands in the colorway, but most of her

Yarn recycled and respun from Ellie's Reclaimed Cashmere.

yarn is worsted weight. To learn more about Ellie's Reclaimed Cashmere visit her website: www.elliesreclaimedcashmere.com

Due to the fact that cashmere and mohair create such long lasting upholstery fabric or garments, it is possible to find vintage pieces in your local thrift store. See the teddy pattern that features fabric from a 1940's era coat. Wouldn't this be a wonderful way to preserve a memory? Imagine a troupe of bears created from great grandmother's mohair or cashmere clothing.

Angoras seem to be less hardy animals when compared to their dairy and meat goat cousins. Special care at birthing time, the provision of shelter during harsh weather, and frequent health checks will be important in the overall mortality rate of the herd.

The reproductive cycle of the fiber breeds is the same as the dairy breeds, with gestation occurring 145 to 150 days after breeding (see chapter 2, page 20). Twins are less frequent than in dairy herds.

In my personal experience, I have found angoras to be attentive mothers, with rejections rare. The kids are the most precious creatures I have ever seen. They are bouncy little balls of curls with eyelashes like that of a china doll. These animals are the cure for the midwinter blues, and their births are something I greatly look forward to each year. Baby goats are truly miniature versions of their adult counterparts: tiny little hooves, little knobs that will become horns, and those tight curls that make them appear they have just had a curly perm. . . . Irresistible.

Cabin Creek Studio

Handcrafted on an
Antique Circular
Sock Knitting Machine

Handmade socks from Cabin Creek Studio. Sue Vunesky knits socks on an almost daily basis using antique Gearhart and Erlbacher Gearhart sock-knitting machines. Sue imports yarn from Germany to make her rainbow of sock designs.

If it has to do with historic arts, it is almost a guarantee that Sue Vunesky has studied and mastered it. Well known for her baskets and handmade brooms, a sock-knitting machine caught her eye several years ago. Sue states. "I was all signed up for a class in sock knitting at a fiber event. Wouldn't you know it, I had to miss the class. That was like waving a red flag in front of a bull." Sue turned to eBay and bought her first sock-knitting machine. "The instruction manual was all in French and it was put together backwards." Sue soon discovered a group online and a member

who lived nearby. "I was thrilled when I got the machine to make a tube. It took two months of trial and error to actually make a sock and another two months to make one that matched. However, once you understand the process it becomes so simple." She can now make up to five pair of socks per day! Sue has made numerous patterns for gloves, mittens, fingerless gloves, even a rabbit all made with the assistance of the sock-knitting machine. Patterns are available through her website, www.cabincreekstudio.com.

Sue, a former machinist, appreciates the technology that went into creating the circular sock-knitting machines. Some of the earliest models date back to the 1800s and production continued until the 1920s and 1930s. Originally made for use in factories, the sock-knitting machines caught on and many were sold for home use. The machines were of good quality but apparently were sold as an early "work at home" business, for which the homemaker was to make socks and sell them back to the respective companies for resale. Apparently, it was a rare event when the socks perfectly met the criteria and were actually purchased from the homemakers. However, the trend of making socks at home with the assistance of a sock-knitting machine caught on and several manufactures made the machines for decades. Legere, Auto Knitter, and Gearhart machines are still available today via eBay and other collectors' sites. There are aftermarket parts available for repairs.

An array of colors will create stripes and solids, making each sock unique. "It takes practice to make two just alike!" says Sue.

Erlbacher Gearhart Sock-Knitting Machines

New sock-knitting machines are available from New Zealand and now from the United States. David "Pee Wee" Erlbacher is manufacturing these fascinating machines in the United States (www.erlbachergearhart.com). Pee Wee is a machinist with a keen eye and an understanding of how things work. His cousin, an avid collector of all things, showed him an antique sock-knitting machine and this intrigued Pee Wee to the point that he bought an old machine on the Internet and started calculating exactly how to reproduce the working mechanism. Pee Wee states, "I went to the school of hard knocks. I worked and figured out how to make the machine but we didn't know how to make a sock. I just liked the way the thing worked. You turn the crank and it spits out a sock!"

Pee Wee thought he could make the machine in a few weeks. In the end it took 19 months to perfect it. "We wouldn't sell them before we tested them and knew they worked perfectly." With the assistance of several top machinists and CAD programming, the programs for creating the parts were hammered out via computer.

Family members Jamie Mayfield and Grayson Erlbacher are now the knitters and test each machine. They, too, became fascinated with the process of sock making. The two women have an innate understanding of how the machine works, which assists them in "cranking" out beautiful handmade socks. Walking through the

Sue Vunesky turns the crank on her sock-knitting machine and counts the number of turns, pulling out needles to create the heel. The machine will create the ribbing too. The yarn carrier travels around, and the latches on the needles close around the yarn and knit it into the piece. This is fascinating to watch. After years of practice, Sue can make a pair of socks in 30 minutes.

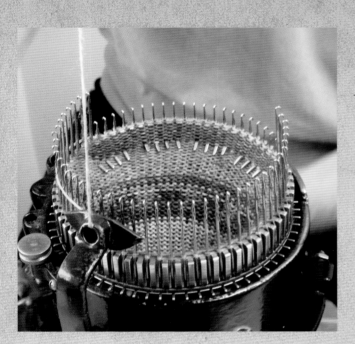

Metal prong forks with attached weights assist in the creation of the heel. Notice the "pocket" forming for the heel in this photo. To make the heel, Sue disables some of the needles so the machine makes only partial revolutions.

machine shop, it is not uncommon to see cones of yarn and projects on the knitting machines. Sales are good, and news regarding the availability of the machines is spreading. Glancing down at Pee Wee's feet, I noticed a bright colored pair of handmade socks peaking through his Crocs. Jamie states, "Pee Wee has a problem! He knits a lot." The family states their goal is to revive interest in the machine and to "bring people to the craft."

The bit of white waste yarn simply serves to start the sock on the machine. This yarn will be removed and the toe closed to finish the sock. Note the colorways in the kitting. To make a second matching sock, the yarn must be started at the same place in the dyed pattern.

Fiber Projects

For the purposes of this book, I have focused on projects that do not require a major investment. In fact, the majority of the projects featured require equipment which will cost $25 or less. Small looms, felting equipment, doll faces, or push molds are all very economical. Of course, mohair and cashmere can create beautiful woven garments, and if you have access to a loom, then by all means experiment with spinning the fibers and creating luxurious garments. If you have your own fiber to work with, so much the better. You may also substitute these fibers as you spin to create wonderful yarns for knitting or crocheting.

I want to introduce to you a few art projects that will showcase the fiber, creating unique projects, perhaps taking you a little outside the typical artist's box. These projects are all starting points, and I'm sure your own imaginations will kick in and expand these projects even further.

✤ Dyeing ✤

Fiber may be dyed at several stages after the initial cleaning. Dry fiber may be added to a dye bath before carding, creating lovely variances in intensity. Wool may be spun, and then dyed in a skein. Mohair particularly picks up dye with trueness to the intended color, a vibrancy and intensity not found with other fibers, and dyeing enhances the natural sheen of mohair.

Experiment with types of dye. You'll find a huge variety of specialty products intended for the spinner. Some have varying shades of color built into them, giving a gradation of color within the fiber. These are my favorites.

In this section, we'll look at methods for dyeing yarn or roving with painting and jar dyeing methods. Natural dyeing is a science unto itself.

Yarn Painting

Yarn (or roving) painting is one method of adding color. This technique works well with long pieces of roving (8 feet long or so) or skeins of yarn. To paint roving, begin by tying lengths of roving into a bundle. Place the bundles in hot water with a squish of dish soap. Let the roving sit in the hot water for about an hour. Push the bundles into the water but do not agitate them (we don't want felt). After soaking, drain the water and push the bundles lightly to remove excess water.

Cover a work table with plastic. Next lay down lengths of plastic wrap long enough to accommodate the wet roving. Tape the ends of the wrap in place. Mix up your dye in plastic containers. Unroll the roving—it should not be dripping wet—and place it on the plastic wrap. Depending upon the width of the roving, you may want to add several pieces (up to four) per piece of plastic wrap. Purchase a large syringe (no needle) or small plastic bottle with a tip. Add the dye to the syringe or bottle, and apply the dye to the roving. Add it in a manner that is pleasing to your eye. Shading can be subtle or bold. Various colorations and color values make for an interesting end result. Think of rainbows, Neapolitan ice cream, autumn pallets, and variances of light blue to indigo with splashes of purple. This is the place to let your inner designer come through! Be sure to make enough to spin up into a full project. There will be no recreating this exact yarn again, so you'll be crafting a one-time only project! Of course, similar yarns can be reproduced but there will always be slight variances (dye lots), which is what makes this project interesting.

Once the desired effect is achieved, remove the tape and roll up the roving in the

Carol Leigh of Hillcreek Fiber Studio at the dye pot. Allowing the yarns to soak for varying periods of time creates a deeper or lighter hue.

plastic wrap, making a packet. It is ok if the ends don't meet in the middle. Steaming is required to set the dye. Using a kitchen canner, add enough water to cover the bottom of the kettle. Place the roving, still in the plastic wrap, in a rack, suspended above the hot water. Remember, you want steam not boiling water. A rice cooker will work for this, too. Allow the roving to steam for about 45 minutes. Make sure the kettle does not boil dry.

Remove the roving from the pot being careful of the steam and hot wool. Allow it to cool to the point where the wool can be handled. Have a sink full of hot water with a squish of dish soap ready, then unroll the roving (throw away the plastic) and place the wool into the hot water. Let it rest for about 15 minutes. Squeeze out the excess water, and

drain the water from the sink. Refill the sink with warm rinse water and add the roving one more time. After 15 minutes, drain the sink and squeeze the excess water from the dyed fibers. Spin them out (no agitation) in the washing machine, and then hang them to dry. The effect before spinning is almost like that of a tied-dyed technique. Skeins of yarn may be dyed in the same manner.

Space Dyeing

Another interesting means of dyeing fiber is space dyeing. Recently I dyed roving in a crock pot! No kidding, a crock pot. Of course, you'll need a designated pot for this method, and it will only work for a small quantity of roving. Used cookers are available at yard sales and thrift stores.

To space dye in a crock pot you'll need four quart-size canning jars, white vinegar to set the dye, and 4 ounces of wool/mohair blend roving. Choose any kind of dye, such as Kool-Aid, Easter egg dye, home dyeing preparations, or fiber dyes.

Begin by preparing the dye bath in the canning jars. Mix a different color in each jar according to the directions. Add ¼ cup of white vinegar for Easter egg dye or Kool Aid only. Add enough water to fill the jar almost to the top. Place the jars in the crock pot.

Divide the roving into four parts; add one part to each jar, leaving a tail. Dip the tail from one jar into the jar next to it and submerge the tip in the dye. This connection will bring the dye from one jar to the other, subtly blending the fibers, one color to the next.

Fill the crock pot with water, almost to the top of the jars, and then turn it on low for 3 hours and add the lid. Turn the crock pot off after 3 hours but let the fiber remain in the covered cooker for 3 more hours. Allow the fiber to cool, then place it in a sink filled with warmish water to rinse. Rinse until the water runs clear. Spin the roving out in the washing machine, being careful not to agitate it. Hang the roving to dry, spin up the yarn, and glean the rewards.

✤ Spinning ✤

To spin, you will need some equipment. A drop spindle requires a modest cash outlay. A very primitive model can be made from a pencil and a potato! If you are serious, though, try one of the inexpensive spindles to make the process go a bit more smoothly. If you are ready to invest in a spinning wheel, visit a fiber shop and try out different models. In the beginning, I found spinning to be quite frustrating.

However, over time, I developed a rhythm and found the process quite restful and relaxing. As with all new skills, it is a process.

Drop Spindle

A good drop spindle is weighted and designed to spin evenly. This ancient design is still very much in use today. Simple in design, the spindle consists of a shaft and a weight. The shaft is used to carry the spun yarn, and the whorl serves as a weight. What is yarn? Simply fibers twisted together. There are several types of spindles: top whorl, bottom whorl, and Navajo. The top whorl is recommended for finer yarns, the bottom for heavier weights, and the Navajo for heavy weight yarns. Spindles are available in light and heavy weights. A medium weight (2 to 3 ounces) is easier to learn on.

These directions are for the bottom whorl. First attach a length of yarn (18 inches long) to the spindle and use that to begin the spinning process. This is the leader. Tie the yarn leader to the whorl, and then carry the yarn underneath the weight to the bottom of the spindle shaft. Loop it around the shaft, and bring it back over the side of the whorl. Carry the leader up through the hook.

Tie the leader on to the shaft. Set the spindle down. Draw out lengths of carded fiber to separate lengths of hair. This is called drafting. Place a piece of carded fiber in both hands, and pull (draft) the fiber until it begins to separate into a long continuous length. Pick the spindle up (if you are right handed) in your right hand. Give the spindle a flick to start the clockwise movement. Practice this to get the hang of it. Once you feel comfortable with the spin, then take the leader yarn and fluff the end of it a bit. Have a rolag nearby,

and set the spindle in motion. Pick up the rolag, and place the end of it with the frayed end of the yarn. The spinning motion will grab the rolag and begin to combine it with the end of the yarn. You are off and running. Continue to spin the spindle, allowing the weight of the device to pull out the fibers from the rolag. When you have used up the supply of yarn, leave a fluffy bit on the end to start another rolag.

After a good length of yarn (yes, it is yarn now) has been spun, spot and wind it onto the spindle. Begin again and continue to spin the rolags into yarn. I think this gets easier as you go along due to the facts that the process becomes more comfortable and that the weight of the spindle increases, drawing

down on the fibers. Control the density of the yarn by pulling out a small or large amount of fibers. When the spindle becomes full, remove the yarn and begin again.

In the beginning, expect variances. At first, the goal is to get the technique down. Consistency will come in time. The first yarns become designer yarns with slubs (thick spots) and thin areas. Ironically, this type of yarn costs quite a bit when it is crafted that way on purpose! Designer yarn can be difficult to knit, due to inconsistencies, but it is perfect for weaving! The project will be enhanced by uneven spindling, and the end result will look as if it was planned that way.

When removing the yarn from the spindle, create a skein by attaching the exposed end of the yarn to a chair back. Work the yarn between the extensions on the back to form a continuous loop. First efforts will be quite curly and have the tendency to loop back upon itself. Working back and forth on the chair will create the skein. To set the twist, tie the skein (while still on the chair) in several places to prevent movement. Fill the sink with warm water and a squish of dish soap. Remove the tied yarn from the chair and submerge it in the water. Let it sit for several minutes, take it out, and roll it up in an old towel to absorb the moisture. Hang it to dry, adding a weight to the bottom if it tends to curl up. Remember: Do not agitate the yarn while it is in the water or you will make felt!

Knitted Scarf

For the simplest scarf ever, cast on 15 stitches on size 13 needles. Check online resources for basic knitting instructions, including casting on.

You will need two skeins of bulky yarn, approximately 207 yards. All that it takes to do this scarf is knitting. This may sound a little too simple. However, when using a bulky yarn, large needles, and a few accents, this is a perfect, quick-moving project.

Add mohair locks into the knitting by holding the tail of a lock and working it into the knitted stitches. Another idea is to use a mohair yarn and then use intact locks for the fringe. This is such a quick and easy project—and it's a good way to use bulky yarns. Perfect for a handmade gift!

This knitted scarf combines a ready-made mohair-and-wool blended yarn with unspun mohair locks. Simply work the locks into the knitting by laying them on top of the yarn while continuing to knit. The locks add texture and retain the curl mohair is known for. Large knitting needles make this a quick and easy project.

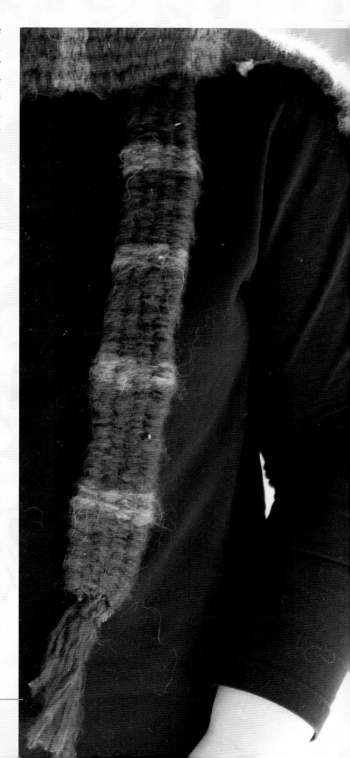

I-cord is very versatile. It can be used as a sash, handles for shoulder bags, or even flashy necklaces when woven with a sparkly, lightweight yarn. When whip stitched together to form a coil, hats and bags can be created from yards and yards of I-cord. This black I-cord is shown with a Knitting Nancy.

I-Cord Pouch

The pouch pictured is worked on a French knitting tool, sometimes called a Knitting Nancy or knitting spool. These tools are readily available at craft supply stores and are under $10. The tools are made of wood or plastic with metal posts inserted in the top. By working the yarn on the posts, the tool creates an I-cord tube, which can then be lashed together to form a coil.

Using the French knitting tool, form the bottom of the bag, and then work in a circular pattern to continue to grow the bottom of the bag. When it is as large as you desire, begin to work up, rather than out. For a small bag, use a 2-liter bottle or other form to help to shape the project. Decide how tall you want the bag to be, then add 3 inches and cut the bottle to that height. Put the bottle in the middle of the coil and begin to work up the bottle. When you reach the top of the bottle, end the coil by sewing it securely in place with needle and thread.

Add handles made of more cording; secure those to the inside of the bag. A loop and button closure makes a cute bag. Size this to carry a water bottle or cell phone, or make it even larger for a shopping bag.

I-Cord Strings

I-cord is also useful for purse straps, belts, necklaces, hoodie strings, and even shoe strings!

Materials

- 1 skein of heavyweight worsted yarn
- Measuring tape
- Scissors

Decide upon the desired length and cut yarn according to the directions described in the I-cord pouch project. Use the French knitting tool as instructed in the I-cord pouch project. Tie off the finished project, add in beads or charms, and enjoy the finished product.

⚘ Weaving ⚘

Simple looms can be constructed from wooden frames designed for the purpose or from a picture frame.

A freeform project requires only a stick from the woods as the form for the weaving. A Styrofoam board helps to keep the weaving taut as you work, and T-pins keep the weaving in place.

I-Cord String

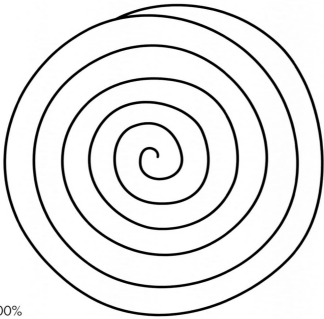

Reproduce at 100%

Details of the freeform weaving. A Styrofoam base and T-pins help hold the warp (vertical) strings in place. Hemp is the fiber of choice for this piece, with a wool weft woven through wool yarn, cotton, jute, or other fibers can be used as the warp or weft. Picture items like feathers, bells, or seashells tied into the weaving. This would be a great way to preserve memories from a particular vacation—the colors and textures can recreate an ocean view or a perfect sunset.

Warp and Weft

In simple terms, the warp is the vertical or horizontal background of the weaving. Usually the warp goes from the bottom of a frame loom to the top to create the structure of the piece. The weft is then the filler. The warp is best when it is a strong fiber, evenly spun. Lumpy or ropey yarn can be difficult to keep in place as a warp, especially for a beginner. A nice, evenly spun mohair or wool yarn, linen, or hemp cords make an interesting framework on which to build your project.

The weft, however, is the place to use those unpredictable designer yarns that are made by accident while learning to spin or by planning by the accomplished spinner. In many projects, the more texture the better. Basically it comes down to the eye of the artist and designer. If you have not applied those terms to your thought process, then it is time to do so. The creative spirit is a powerful force, and thinking in those terms will strengthen every project. Weaving can be as careful, planned, and geometric as you choose or an avant garde affair where anything goes. Let your artistic eye be your guide.

Mohair is a slick fiber, and this attribute is what makes the beautiful sheen and fuzzy texture. Those same attributes can make it somewhat challenging to spin. Carding mohair with wool can offer the best of both fibers, and this is a widely used practice. Wool is also less expensive than mohair, and blending the two will make the mohair go further while still providing the desired effect. If you are a purist and do not wish to combine the fibers, this is not a problem. Once you become accustomed to spinning one or the other, it will become second nature.

Unconstructed Weaving

This freeform weaving highlights natural elements. Find a stick, or for a more refined base, use a dowel rod. The rod can be painted to add more color to the piece. Simply let your eye be your guide and tie on linen warp threads in sections. This weaving has five warp threads per strip.

For weavers you can use textured yarn or even mohair locks. For ease of weaving, use T-pins and a piece of Styrofoam to anchor the project while you weave. Then proceed, over and under, paying close attention to the edges (the selvages). This is where weaving can go awry. The tendency is to draw the weft too tight or to make it too loose. Keep this in mind in any weaving project. I usually weave with a ruler nearby to check my piece, which should be the same width from top to bottom and not get skinny in the middle.

Once the design suits you and the weaving is complete, tie in beads, feathers, or other natural elements to complete the piece. Hang it where it will catch a breeze and allow the air to move through it. A larger version of this piece makes an interesting project to place in a window.

Circle-Frame Weaving

Another option for freeform weaving is a circle frame, such as an embroidery hoop. Circular weavings are sometimes referred to as mandalas. The translation of this word from Sanskrit is "magic circle." This type of weaving really does not follow a pattern.

To begin, decide upon a starting point. Wrap the ring several times with the yarn of choice. Move the yarn across the circle (a metal ring in this project), and loop it around

This weaving reminds me of a free-flowing, colorful piece from the 1970s. When hung in a window, it catches the light and flutters in a breeze. One of my planned projects for the future is a large-scale piece, made in the same order, to create a spectacular one-of-a-kind window curtain.

131

the other side. Make a few loops then travel back. Tie off the yarn, and begin with another color. Make a few loops, travel to the other side, make a few loops, and then come back. Quickly a pattern will start to emerge.

There is no right or wrong way to do this type of weaving, and everyone who attempts to make a circular weaving will have a different outcome. Something about this type of weaving reminds me of the God's eyes made a long time ago in Girl Scouts. However, the circle invites the use of the creative touch. Note the long fringe, which adds drama and movement. Work in beads, little pieces of mirror, shells, and other embellishments if desired.

I often look to nature for inspiration for my weaving projects. I sometimes leave bits of yarn or pieces of fiber resting in the fork of a tree. I later find these colorful bits in birds' nests and mouse nests out in the field. Baltimore orioles, however, get the prize for the most incredible nests I have ever seen. How do they know how to pick the right branch on which to build? The orioles construct a bag that is looped over a branch, allowing the wind to move the pouch without blowing it down. Nature is phenomenal.

Stick Weaving

This type of weaving has been practiced for centuries. It is quick moving, easy enough to do with children, and can be used to create a number of interesting projects such as belts, sashes, headbands, bracelets, scarves, and bags.

You can buy stick-weaving kits or make your own from dowel rods. The rods can be any size, depending on how large (and dense) the project is. Note the length of the dowels, the pointed ends, and the flat ends, which have a hole for the yarn to go through. To make your own sticks, purchase dowel rods, and using a pencil sharpener, sharpen the ends. The ends should not be too sharp, so after using the pencil sharpener sand the points just a bit. Then, using sandpaper, sand off a portion of the opposite end of the dowel to make it flat. Mark the area with the pencil, then sand until the indentation is created. Drill a small hole. Repeat this process for each stick.

Decide upon the project, then cut the warp yarn to thread through the sticks. The warp yarn will be double, so if the desired strip is to be 24 inches long, double that measurement to 48; then, add 2 to 4 inches on each end for tying and fringe. Thread each stick with the yarn and tie a knot in the ends. Repeat for each stick. Hold all six sticks in one hand, and tie the weft yarn to the first stick. Begin weaving over and under, keeping the sticks evenly spaced in your opposite hand. The weaving will become very easy after the first two or three rows are in place. Simply continue weaving back and forth, over and under the sticks. Remember to pay attention to the edges (selvages) to keep the tension of the weaving uniform. Weave until about 4 inches of the warp yarn is left, then cut it free from the sticks. Tie it off to prevent unraveling. For a sash or a belt, consider leaving the fringe longer and working in some beads for effect.

Stick Woven Posey Doll

This little figure is perfect to attach to a card or a pin, or create a series of them and place them in a hand-woven pouch for a pocket full of poseys. This doll is about 5 inches when finished.

Materials
- 1 set of 6 stick-weaving needles
- Jute for body (You may also use hemp, linen, or cotton twine.)
- Yarn for weaving (Mohair blends are nice.)
- Beads for decorating
- Pin back if desired

Cut six pieces of jute 15 inches long. Thread one piece through each needle, and knot it at the bottom.

Tie the weaving yarn to Stick 1, then pick up Stick 2 and hold them both in your left hand. Weave back and forth, under and over, looping around the outside stick. Keep the sticks very close together, weaving until you have 1½ inches of weaving. Measure 4 inches additional yarn, and cut your work free from the yarn ball. Set it aside.

Pick up Stick 3, and tie on the weaving yarn. Weave around Sticks 3 and 4 as above until you have 1½ inches of weaving completed. When this piece is the same size as the first piece, pick up Sticks 1 and 2, holding all four in your left hand. Holding the yarn attached to the ball and the 4-inch tail together, weave across all four sticks. Continue weaving until the tail is worked in; then, keep on going with the single strand until the body is 2 inches long. Remember to keep the sticks close together. Work the weaving down but not off the sticks as the weaving progresses. Keep the yarn taut to retain the tension.

When the body is complete, weave the right arm using the yarn still attached to the ball. Pick up the Stick 5 and place it on the outside (right) of the weaving. Weave the arm, back and forth over two sticks. At 1½ inches cut the yarn and tie off the weaving on the outermost stick. Tie it securely so the weaving won't come undone.

Weave the center section for the head from jute. Tie the jute on Stick 4 and weave Sticks 4 and 3 with 1 inch of weaving. Tie off your work with a knot in the back.

Pick up the Stick 6. Introduce a new piece of yarn, and tie it to Stick 6. Place Stick 6 beside Stick 5 and hold all six sticks in your left hand. Weave 1½ inches on Sticks 6 and 5 to make the left arm. Cut the yarn, leaving a tail, and tie off on Stick 6. The entire weaving is now on the sticks.

Now for the fun part. Slide the weaving down onto the jute warp, pulling one stick through at a time. Slide/pull all the way down to the knotted ends to form the feet. Tie the ends from Sticks 6 and 5 together to form the left arm. Tie the ends from Sticks 4 and 3 together to form the top of the head. Leave these a little longer to resemble dreadlocks! Add some beads later. Tie off Sticks 2 and 1 to form the right arm.

Tuck the remaining knots from the arms behind the body, and stick them down with a drop of hot glue to hide them. Add a pin back if you would like to make a pin. Make several to tuck in a pocket, or use them on mittens, slippers—wherever there is a need for a pick-me-up!

Sick Woven Posey Doll

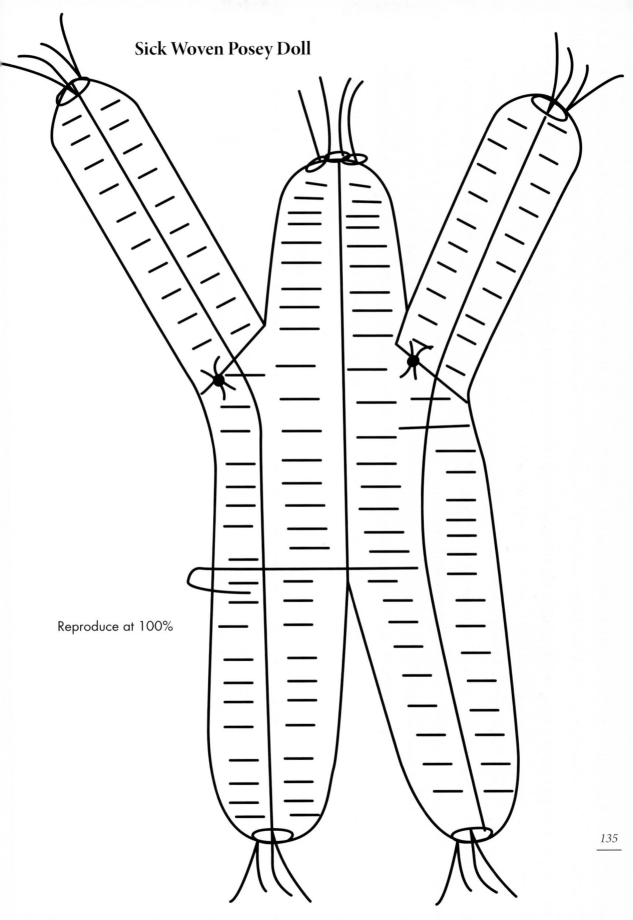

Reproduce at 100%

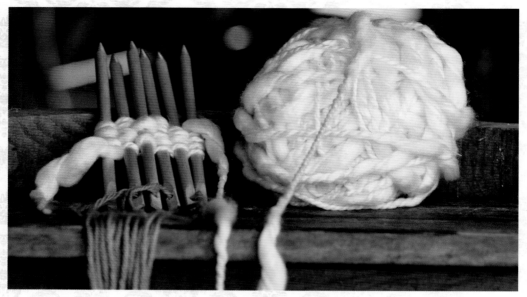

Stick weaving is an ancient art form. Weaving sticks are easily made out of dowel rods. Rods should be worked to a dull point on one end and tapered on the other, with a small hole drilled on the dull end to carry the weaving fiber. Working across the green threads (warp) creates a small unconstructed type of loom. The white mohair blend yarn will be used as the weaver and create a dense ribbon-style sash. The sash can be made larger or smaller, depending on the number of sticks used and the diameter of the dowels used for weaving. Bits of the green weft yarn will show through. Additional colors can be tied into the knots at the beginning and end of the weaving. Beads, bells, or other accessories can also be tied into the beginning and ending fringe.

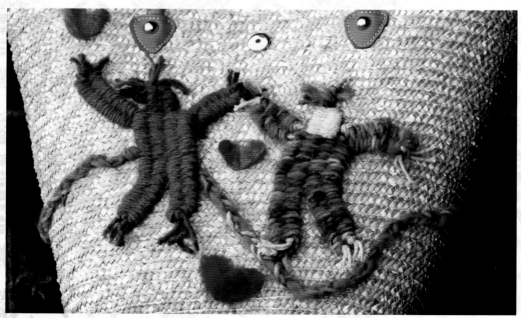

Stick-woven figures adorn an inexpensive woven bag. These happy little creatures, Posies, are the perfect embellishments for bags, boxes, and even cards and journals. Adjust the size by using larger or smaller sets of weaving sticks.

A collection of small looms. The daisy loom (center) makes perfect fun flowers to embellish your knitting, weaving, or sewing projects. A Knitting Nancy (the purple tool) is an older concept once used to teach children to knit. The modern-day version here produces I-cord in two sizes.

Little Looms

If you are of my vintage, then you may recall daisy looms. These are inexpensive plastic looms that make daisies or petaled flowers. Once upon a time, these daisies were crocheted together to make afghans or vests. Recently reintroduced, the daisy flowers are now used on hat bands and purses and show up on clothing as colorful accents.

Another type of loom is a vintage square loom called a Weave It. These are no longer produced but are found on Internet auction sites. The Weave It loom will make a square which can then be sewn together to form fabric for purses, vests, placemats, and other accessories. A Weave It is very portable, a good way to take your weaving with you wherever you may roam.

A Lacis Walking Stick loom is a larger version of stick weaving. With this small loom, a series of sticks are held in a forklike device. This arrangement simplifies the process, keeping all the sticks in order. In this type of weaving, the warp is covered by the weft and only the weft is visible. Hence, the term for this type of weaving is "weft-faced" or "tapestry weave." Weaving in this manner will produce a panel that can be joined together by lashing the edges or using a darning needle. Panels can be used to make saddle blankets, pillows, wall hangings, and other works of art. Mohair is wonderful for projects,

The full-size version of a triangle loom doesn't fit the description of a "little loom," but it does come in a smaller version for laptop weaving. Evenly set nails provide the proper spacing for triangle weaving.
Courtesy Hillcreek Fiber Studio

such as saddle blankets, that will receive a lot of wear.

The full-size version of a triangle loom doesn't fit the description of little looms, but these beautiful looms do come in a smaller version for laptop weaving. The triangle loom is described as "the only loom that warps itself" by master craftsman Carol Leigh Brack-Kaiser. The triangle loom provides the perfect format for the creation of shawls, and nothing is lovelier than a mohair shawl done in vibrant colors. These looms are an investment. They are also heirlooms to pass from one generation to the next. Walnut, cherry, and other fine woods are used to make the loom a work of art in itself. Stands are available to match the wood used in the loom. When this element is standing in the corner of a family room, it is impossible to walk by and not be drawn in to the weaving.

A triangle shawl in progress on an adjustable frame loom. Note the warp and weft—this loom warps itself! Courtesy Hillcreek Fiber Studio

An example of a tartan plaid scarf woven on the adjustable frame loom. There is no end to the versatility of the garments that can be made this loom.

Carol Leigh Brack-Kaiser

Carol Leigh Brack-Kaiser of Hillcreek Fiber Studio is a born educator. Though she does not teach in a traditional classroom and her students range from age 5 to 75, Carol has a gift for sharing her knowledge of fiber arts. When she began weaving on a potholder loom at age 5, she could not have imagined where her love of the traditional arts of natural dyeing, spinning, and weaving would take her. An artist, author, and inventor, she is one part Earth mother, one part alchemist.

As with many effective teachers, Carol remains a student, frequently participating in workshops presented by other experts in the world of fiber arts. During her study for her master's degree in textile and apparel management, her research focused on natural dyeing, specifically on the native plant, poke. The berries have long been known for their dye properties. Carol Leigh has become known as an expert on the subject. "The variance of color in the poke berry dyes are endless. The range extends from red to fuchsia depending on the process." Carol Leigh's research led her back through time to early techniques and another way of life. Participating in fairs and festivals, many in a rendezvous setting, her reverence for the past is evident.

The rendezvous hold great interest for Carol. "It is amazing how an area can be transformed. Picture a sea of white tents, a village of craftsmen. You can find blacksmiths, silversmiths, coppersmiths, and many others who present the work and skills of the time period from 1640 to 1840. Each artisan is a master of his or her craft. You won't see a plastic cup or disposable container. Every detail of the camp is to be authentic to the time period. Of course, my work, my shawls fit right in. Everyone needs a shawl." Carol Leigh travels throughout the United States, teaching and participating in the rendezvous events. She states her favorite moments are in the evening when the traditional instruments strike up a tune and transport her to a different time and place. "Singing, dancing, firelight, and lantern glow. . . . It's magical."

Carol Leigh is best known for her triangle looms, which provide the structure for her triangle shawls. When she began weaving in earnest in 1982, she soon discovered the need for an adjustable loom, one piece of equipment that could make a variety of sizes. Collaborating with her son, Carl Spriggs, the pair soon came up with an original design, which they patented. The unique design provides the structure

Carol Leigh with a work in progress on her adjustable frame loom. Carol's book, Continuous Strand Weaving Method. *represents years of practical knowledge and study.* Courtesy Hillcreek Fiber Studio

Carol Leigh models one of her creations. Various fibers and weights can produce a warm winter garment or a light spring wrap. The fiber choice will create a utilitarian shawl for daily use and quick wrap for trips to the barn or an elegant evening shawl. A dedicated weaver can create a shawl in 12 to 15 hours of weaving.

for the weaving, and it is only loom that "warps itself." Anyone familiar with weaving at all will appreciate that statement. Warping is normally a tedious process. When weaving on the triangle loom, the weaver is almost immediately gratified as the progress is quickly evident. It doesn't take long for patterns and structures to emerge in the piece. The concept of the loom has existed since medieval times. However, Carol Leigh's and Carl's improvements have brought this style of weaving right into the present millennium. "Carl has numbered each one and kept track. We have sold thousands of them now." Due to customer requests, the two inventors created two additional looms, the 5-foot square and 7-foot rectangle. These frames allow the creation of myriad sizes, all employing the continuous strand weaving technique.

Carol Leigh's students are from greatly varied backgrounds. "We recently had a group of nuns. They camped out in the backyard and had a great time." After 25 years of teaching, she states it never gets old. "People come here to destress. We have surgeons, teachers, nuns, and nurses, people from all walks of life." Weaving and working with natural fibers and plants tends to take class participants to a different place, a better place.

Weaving on a Frame Loom

You can use a picture frame to make a loom! Look for a sturdy frame at a yard sale or thrift store. If the frame is wobbly, purchase four small L brackets and screw them into each corner of the frame. This hardware will sturdy the frame for weaving. If at all possible, use a saw and make some small grooves about ½ inch apart on the top and bottom of the frame to hold the yarn in place. The great things about using picture frames is that first, they are very reasonable, and second, they come in such a variety of sizes that you can scale your frame to your project.

To warp the loom, begin on the left side of the frame loom and tie a piece of linen, cotton, or wool thread for the warp. Wrap the frame, placing the thread into each slot. Tie off on the end. Place a piece of cardboard in the middle of the warp so it is clear that we are weaving only on the front of the loom. The rest of the warp, on the back side of the loom, will be used to tie off the piece and for a fringe.

Weaving on a frame loom can be very orderly if evenly spun yarn is used. Note the mohair purse made by the Elelloang Basali Coop (see page 114). By overspinning the mohair, tight loops form to resemble the mohair locks. Overspinning is just as you would think: spinning the mohair very tightly until the loops form.

If an art piece is what you desire, this is the time to use your beginning yarn. Those ropey, uneven pieces are great to use, and it will look as though you planned it that way all along. Work in flashy yarns, metallics, ribbons, or other textured materials to provide depth to the piece. It is also possible to use locks of mohair or carded rolags and simply

beat those into the weaving. A wooden beater (made for this purpose) or a hair pick are good tools to beat the fiber.

To weave on a frame loom, begin by working a piece of cotton, linen, or other sturdy thread for the first several rows of weaving. This is the header. Simply weave over and under the strands of warp. At the end of the first row, wrap around the last piece of warp and start back again. Row 1 should be woven on warp strings 1, 3, 5, 7, 9, etc.; Row 2 on 2, 4,

6, 8, 10, etc.; Row 3 on 1, 3, 5, 7, 9, etc. Weave six rows with the thread. This is a basic tabby weave. (You can add more complicated weaves when the basic process is accomplished.)

Finish the header after Row 6, leaving the thread on the right side of the loom. Begin to weave with yarn now, working the tail of the thread in 2 to 3 inches with the first part of the yarn. This ends the linen and works it in with the new yarn, giving a seamless beginning and end.

Continue to weave one row over and under. Watch the selvages (edges) to keep them even, then travel back in the other direction, going under and over the opposite warp strings. A common problem in any type of weaving is shrinkage, in that the woven piece tends to get smaller in the middle. This is a tension problem. To combat this issue, weave with a ruler at hand and check your piece frequently to make sure that it is not getting smaller as you go. If it is, rip out the weaving back to the point to where it started losing the width and continue on.

To add in a new length of yarn, add 2 inches on top of the tail of the existing yarn. Work those two together, as one, until the old yarn is gone and only the new yarn continues. Make sure to beat the yarns down so there won't be a bump at the point of the splice. Continue to weave the entire length of the loom in this manner. Yarn colors may be changed to add interest using the splicing technique as above.

To finish the piece, go back to the linen or cotton thread and weave six rows again to make the header and footer of the piece uniform.

Woven Bag

To make a bag from a piece of fabric woven on a frame loom, measure the piece and mark it in thirds, using a straight pin for marking. Seam the edge up to the second pin, using a sewing machine or whip stitching the edges together. Press the remaining one third down to form a flap. Braid a short length of fiber to make the loop and sew on a button for closure. Add a length of I-cord to one slide to form a carrying strap. Stitch the strap in place firmly. Press the entire piece to set the shape, using a hot iron. Cover the piece with a cotton towel to avoid scorching the fiber.

Note: The finished size of the bag pictured is 8 x 9 inches before folding it over to form the purse. A 12 x 15-inch loom (approximate size) would be required for this size weaving.

If you would like to purchase a frame loom, rather than using a picture frame, there are ready made looms with accessories. My favorite is from Take It or Weave It. These beautiful looms will become family heirlooms. Complete with handmade tools,

available in two sizes and other decorative variations, this loom makes weaving totally portable. www.takeitorweaveit.com

Think of all the projects: journal covers, bookmarks, purses, larger bags, wall hangings, and more!

✤ Felting ✤

Traditional felting has been around for centuries. Nomadic people made yurts (huts) out of felted llama, sheep, or goat hair. Known for its insulating qualities and durability, felt is a very versatile fabric.

Mohair and wool felt in a slightly different manner. The shiny texture of mohair makes it a bit more difficult to felt than wool. However, once again, blending the mohair with wool combines attributes of both fibers. Felt is also a good way to use up hair from older goats. Due to the change in the quality of the fiber as a goat ages, this is one way to use "grandma's" goat hair. Another is to simply use it for stuffing.

Traditional felting requires carded fibers (see chapter 6, page 108). Beginning carders sometimes purchase dog brushes with metal teeth. These don't hold up for too long, but if you just want to see what carding is all about you could give them a try. Fiber intended for spinning requires carding, too. So the initial investment in good carders will be used throughout your fiber work.

Sometimes wool is not washed before carding. Yarn spun from unwashed wool is referred to as being "spun in the grease." The grease is lanolin, a natural oil found in sheep hair but not in goat hair. Garments made from unwashed wool are water resistant. For our purposes, both the wool and mohair should be skirted (picked clean and all debris

This felt began as wool and mohair blended together with hand carders. Small amounts of wool and mohair are placed on the carders, then one carder is worked against the other. This action lines up all of the fibers. From there, the sheets of fiber are removed from the carders and layered to build a batt. The batt is then drenched in hot water mixed with dish detergent. An agitation process then tightens the individual fibers into a cohesive fabric.

removed) then washed and dried before carding. To blend the fibers together, pick up a handful of wool and lay it across the wool carder with the fibers pointing to the top and bottom of the carder. Lay in the locks of mohair between the wool. Hold the loaded carder in your dominant hand then pass the opposite carder over the fibers. Comb through the fibers a few times, moving both carders in opposite directions. This step simply aligns the fibers and gets them all going in the same direction. After the fibers have been worked back and forth to achieve a smooth texture, gently roll the fibers to

the front of one carder. Continue rolling to create a tubelike shape. This is a rolag. A series of rolags will be required to make a piece of felt. For a small piece of felt create about 24 rolags.

The next step is to wet the rolags with water. Lay a piece of plastic on a table to protect the surface and provide a slippery work place, which is good for working felt. Place the rolags on the plastic. Lay the first layer in one direction to form a 12 x 12-inch square. Go back and lay the second layer in the opposite direction. Mix a small amount of liquid dish soap with a quart of hot water.

Pour half of the hot, soapy water over the wool. Begin to pat the wool with your hands. As the water soaks in, gather up the sides of the plastic and roll it up. Unroll; reposition the now matted fibers, and work it a bit more with your hands. Roll it up again, pat the roll, massaging the exterior surface. Unroll it, and check the progress. If there are any thin spots, add another rolag and work it in. When the desired thickness is achieved, place the felted mat on a screen to dry. Use felt to make any number of projects: purses, bags, booties, hats, and mittens, to name a few.

Needle Felting

Felt can also be made by using barbed needles to work the fibers into themselves. By punching wool or mohair up and down with needles, the fibers become intertwined. To needle felt, you'll need the following equipment: felting needles (singles or a felting tool, which has several needles in a holder); a work surface (such as a piece of foam to hold the work); and mohair, wool, or other fiber.

Lay out pieces of carded mohair, wool, and/or other fibers on the work surface. Using a felting tool in an up-and-down motion, punch the fibers into each other. Repeat the punching motion until the fibers have formed a mat. This is felt. I like this way of felting as it is fast, creates no mess, and is easy to see one texture builds upon another when layering in colors. When using needles to felt, it is important to remember to pay attention. After being stabbed a couple of times by a felting tool (five needles at a time), this point will be driven home! We all must suffer for our art. Do be careful and pay close attention when needle felting.

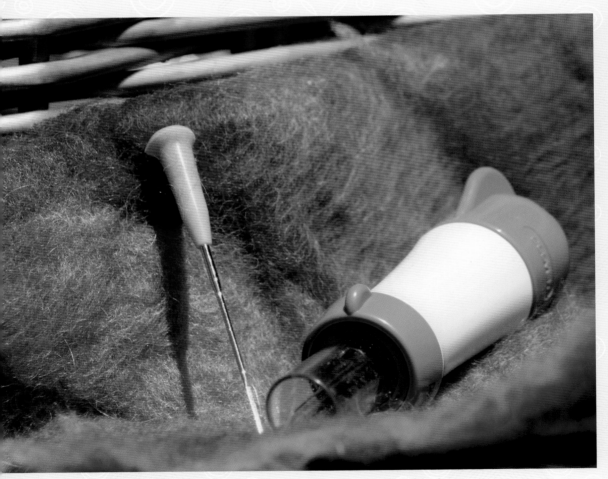

Single needle and multi-needle felting tools. Use the single needle tool for fine details or small areas and the large tool on larger pieces. Each needle has a barb on the end that hooks the fiber onto itself or into a base fabric (see the berets and jackets, pages 150 to 155). Needle felting adds detail, color, and texture to ordinary pieces, creating one-of-a-kind designer items.

This gentle fellow is made of three balls of wool and mohair blended together and shaped into graduated-size balls. After the balls are shaped, use a single needle felting tool and work the fibers into a semi-solid mass by punching the fibers into themselves with the felting needle. This creates a sturdy ball. Add sticks for arms and a piece of polymer clay shaped into a carrot and baked according to directions. Add a little felt hat, and he is ready for winter.

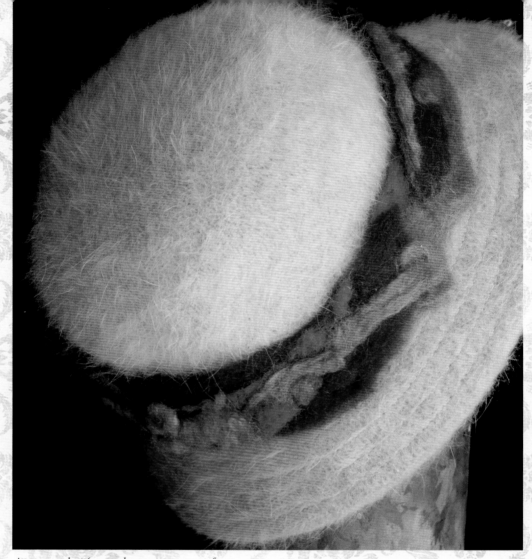

An angora hat (remember, angora comes from rabbits) with a mohair/wool felted hat band. Mohair locks are worked into to provide texture and contrast. The band begins with a solid piece of felt, then embellishments are added one by one. The band is then needle felted onto the hat.

Hat Projects

Being a hat person, I am always looking for a cute chapeau. I found this pink angora blend in a thrift store. Remember, angora comes from rabbits. I liked the color and thought I would embellish the hat with a band. To create this band, I used carded and dyed wool, accented with mohair.

First create the band of purple and teal blue by working the surface with a felting tool, up and down, as described above. Then go back and add bits of the mohair curls in the magenta hue. The textures and colors perfectly complement the hat. When the band is finished, add it to the hat. The band pictured was worked directly into the hat. It is not glued, simply joined to the hat through the felting process.

Beret. The beret is also needle felted with the fiber worked directly into the hat. The fiber literally becomes a part of the fabric of the original piece. Felting has unlimited applications. Picture a wool coat with a free-form art piece felted into place. Purses, bags, scarves, and other accessories can be personalized with bits of waste yarns, beads, and other accents.

A beret is a perfect palette for needle felting. Simple lady bugs are formed from pieces of red wool/ mohair felt. The felt is placed on the beret, then a needle felting tool is used to work the fibers into the beret, creating a permanent bond between the two. Antennae are mohair curls. A running stitch done with embroidery floss creates a bit of contrast.

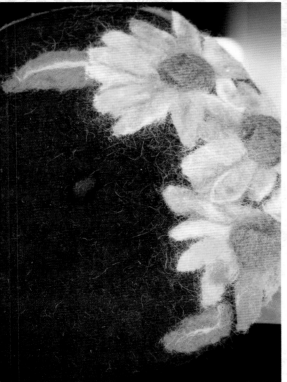

Pieces of wool and locks of mohair form the petals and leaves of this sunflower beret. One by one, each piece is needle felted into place. The design becomes one with the fabric of the beret.

A ready-made red felt hat sports a handmade hat band. The black band is a piece of handmade felt. The floral embellishments are added one at a time, using a piece of upholstery foam as a work surface. See the pattern for vines and flowers on page 154. When you are satisfied with the design, use a single needle felting tool to attach the hat band to the hat.

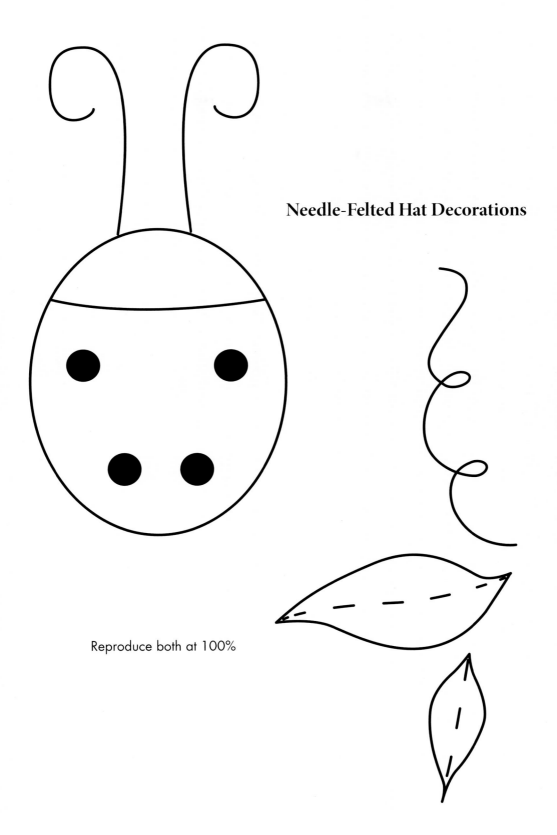

Needle-Felted Hat Decorations

Reproduce both at 100%

Needle-Felted Hat Decorations

Reproduce both at 125%

Wool Jacket

Look for a vintage wool jacket at your favorite thrift store. Hang the jacket to get a feel for the lines. Get in touch with your inner designer, and go to work. Create felted cuffs and pocket accents for a subtle effect. For a true piece of wearable art, work out a design using leftover bits of yarn, then add some needle felting, ribbon, beads, or whatever strikes your fancy. Once again, there is no real pattern for this type of piece. Your finished piece will be one of a kind.

Needle-Felted Figures

For this type of needle felting, use single felting needles to secure the fibers to each other. To form small figures, such as a rabbit, take a look at a photo and see how the rabbit is shaped. It has a pear shaped torso, an angular face, paws that extend from hocks, and of course, ears and a fluffy tail. To make the rabbit, first form the body from wool and mohair blended together. Make three rolags and roll them up into a ball. Then, using a single felting needle, begin to jab at the fiber. You will see the fibers tucking into each as they form a solid ball. Keep working the ball, moving it slightly until a uniform mat has been achieved. Leave a small piece on one end unworked.

Shape that piece for the head, looking at your photo as reference. Work the head as you did the body, again leaving a small piece of fiber loose at the back of the head. When the head is thoroughly felted, place it on top of the body and join the two unfelted pieces by working them with the needle. If the joint seems weak, add a little more fiber and work it until the joint is solid.

Add two small worked pieces for front legs. These legs can be worked until they

Paula Dace models a freeform needle-felted design on the back of a thrift-store jacket. Wool and mohair were felted together to form a batt for this design. A needle felting tool was then used to join the felt to the jacket. Small glass beads were worked into the veins of the leaves.

155

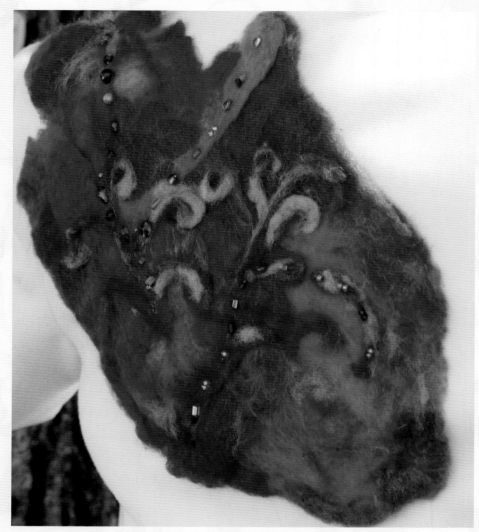

The detail on this jacket reminds me of Mardi Gras. First a felt batt was created then attached to the jacket using single-needle felting. Small plumes and swirls were then worked on top of the batt. Glass beads add a touch of glam and were sewed on with a running stitch.

adhere to the body, or using a long needle and thread, simply pierce the leg and body and stitch them into place. Use the same procedure for the ears and tail.

Add whiskers using lengths of embroidery floss. Using a needle and two strands of black floss, tie a knot at the end of the floss, leaving about a ½-inch tail. Pass the needle through the nose area, and pull it through to the other side. The knot will catch on one side, creating the whisker. Tie a knot in the other side, and leave about ½ inch remaining for the

other side of the whisker. Add two or three of these for the bunny whiskers. Tiny seed beads are stitched in place for the eyes. A bit of blush is added to make rosy little cheeks.

Have an idea of the shape you are going for; then, work the fiber, tucking in one area at a time until the piece resembles the body part you are working toward. This type of sculpting takes some practice. If you would like your animal to stand, cover pipe cleaners with fiber and make legs to support the body. It is difficult to work the wool over

Use single felting needles to secure the fibers to each other to form small figures, such as this rabbit. See how he's shaped. He has a pear shaped torso, an angular face, paws that extend from hocks, and, of course, ears and a fluffy tail.

the pipe cleaners. The best approach for this is to pull out lengths of carded fiber and wrap the pipe cleaners. A dot of glue will help to hold the end in place. The pipe cleaner will go up into the inside of the felted body; just make a small hole with a sharp object, then force the leg into the opening. It may be necessary to work the fiber a little more after this insertion to anchor the legs in place. A drop of glue added to the pipe cleaner before the insertion will also make sure the legs stay in place. I find liquid white glue works well for this step.

Cat Variation
To make a cat, adjust the size of the ears. This cat is in a lying position. It would be easy to adapt the bunny directions to make a standing cat. The best way to make the animals is to look at a picture and study their joints, where the legs attach, and the shape of the head.

Recycled Sweater Projects
Wool, mohair, and cashmere are such durable fibers that when visiting a thrift store, it is not difficult to find a collection in one trip

Needle felting brings fiber to a new level. This kitty began as a simple oval shape for the body. Using a single needle felting tool, the fibers are worked in and out with the needle, creating a sturdy felt body. Leaving a little piece unworked at the neck allows the addition of the head, another oval worked then attached to the unworked extension of the body. Look at a picture of an animal to get the best placement for facial features and details.

down the racks. If you wait until the middle of July, it is quite possible to locate wool, mohair, and cashmere sweaters for $1 each. What I am going to suggest is in direct conflict with the normal care of a natural fiber garment. Our goal is to make felt from a recycled garment.

First, group colors together and only wash garments of the same color family

together. Put the sweaters in the washing machine, and fill the machine with hot water and a squish of dish soap—not too much or you will be participating in a scene out of a Lucille Ball show! Let the sweaters go through the entire cycle. The heat and the agitation from the washing will shrink the fibers together. When the cycle is finished, remove

These decorated pink boots are perfect for the style-conscious 5-year-old. Purchase ready-made pink boots and silk butterflies from a craft supply store. The butterflies in this example had a foam body. I left that intact and simply wrapped a length of fiber (mohair/wool blended) tightly over the body to make the segments of the insect. The ends are hot-glued in place.

the sweaters and put them in the dryer on high heat. You may be quite surprised at the result when the sweaters are removed from the dryer. A man's wool sweater would now fit a 5-year-old boy. (I have had this happen before by total accident! I had no idea that I could have made something wonderful out of a disaster).

Note the density of the fabric now. It is thick, almost padded, and guess what? It is already felted and ready to recycle into amazing hats, scarves, mittens, slippers, and more. The only limit to this method is the amount of fabric in the garment. To get around this limit, combine complementary colors if more fabric (felt) is required to make a finished product.

These mittens lived their first life as a man's sweater. After washing the wool sweater in hot water and placing it in the dryer, it shrank to about 25% of its original size. The fibers matted together and formed a beautiful, dense fabric. Using the pattern provided, simply cut out the mitten shape, using the existing ribbing for the cuff. The mittens are warm and a great way to recycle a sweater that is past its prime.

Felt Mittens from a Recycled Sweater

Mittens are too simple. Look at the sleeve of the felted sweater. What do you see? Hmmm . . . a cuff, two sleeves, two cuffs. Trace around your hand (or the person you would like to make mittens for) and make a pattern. Place the pattern on the sleeve, incorporating the cuff into the design. Cut out mitten one, then mitten two. Even if you are not a seamstress this is not difficult. All that is left is to add felted embellishments, machine or hand embroidery, buttons, and beads and you will have a beautiful pair of mittens.

Lay out the mitten, and add your decorations. When you are happy with the design, attach the embellishments, embroider any leaves, flowers, or other patterns. Once the decorating is complete, turn the mittens wrong side out and stitch the front and back together on a sewing machine using a ⅝-inch seam allowance. Sew around the open area, and trim off the excess fabric to just above the seam. Turn the mittens right side out, and press them with a hot iron, if needed, to make the seam lay down.

These mittens are perfect gifts. What else could you give that would be as pleasing as a pair of handmade mittens—and it's a recycled fabric project with the raw materials coming in at $1. Guess what? There is more fabric in the body of that same sweater to make a coordinating hat!

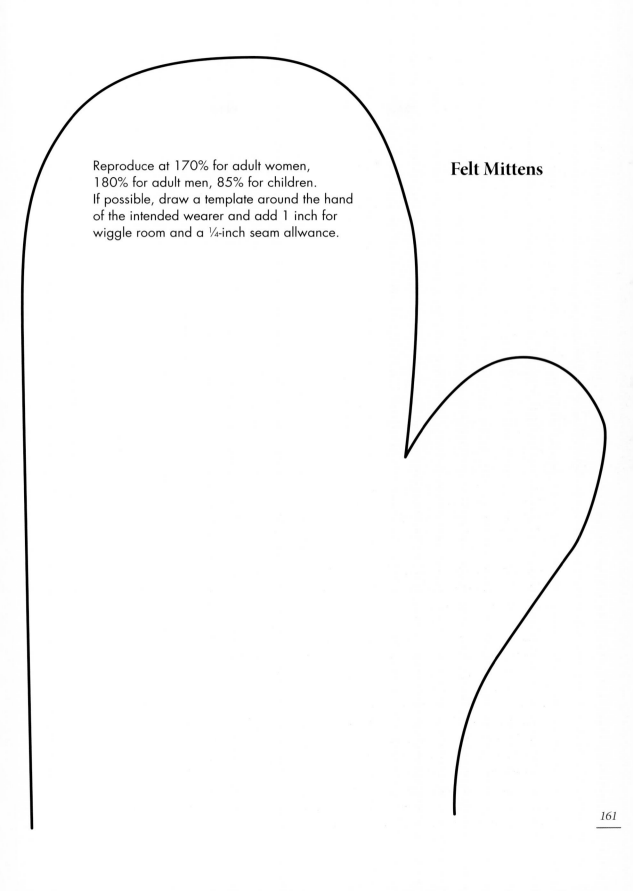

Reproduce at 170% for adult women,
180% for adult men, 85% for children.
If possible, draw a template around the hand
of the intended wearer and add 1 inch for
wiggle room and a ¼-inch seam allwance.

Felt Mittens

Felt Hat from a Recycled Sweater

Take a look at a simple stocking cap to get an idea of how easy these are to make. The circumference of the hat for this pattern is 16 inches finished. Enlarge or reduce the pattern depending upon the wearer of the hat. This hat will fit an adult. Place the pattern on the felted sweater around the ribbed bottom band. This will be the bottom of the hat. Pin the pattern in place and cut two pieces (front and back). Cut through both the front and back of the sweater at the same time so you end up with two pieces the same size. Add decorations such as daisies woven on a loom, felted pieces or buttons, and decorative stitching. Using a sewing machine, sew the pieces together using a ⅝-inch seam allowance. Trim the edges of the seam to ¼ inch. Turn the hat right side out and press the seam in place.

Teddy Bear

Everyone needs a teddy bear or two. Tuck one in a basket or on a shelf to add a touch of childhood whimsy. Make a collection out of a cherished piece of fabric or from a vintage coat.

The bear pictured is made from a thrift-store coat—a wool and mohair blend. To use the fabric effectively, take the coat apart. Remove the sleeves, buttons, lining, etc. Often the lining in the old coats is satin, so save that for another project. Keep the buttons to start your button collection.

The pattern shown is quite simple: a body/head piece, ears, legs, and arms. This pattern is simple and not too complicated, even for a beginning bear maker. Lay the body/head piece on a doubled piece of fabric and cut both pieces at the same time. If the fabric is too heavy, cut the body/head pattern out on a piece of cardboard, lay it on the fabric and trace around it. Cut two pieces alike. Cut four ears, four legs, and four arms.

Place the pieces right sides together. Using a sewing machine (preferred) or whip stitch, sew a ¼-inch seam around the head/body. Sew two of the ear pieces together to form one ear. Sew the two other ear pieces to form the other ear. Turn the ears right side out, put a very small amount of stuffing inside. Now turn the body/head right side out. Position the ears and using a heavy duty needle and thread, whip stitch them into place (see ear placement dots on pattern, page 165). Give them a little tug, if needed to straighten them out. Stuff the body/head firmly using fiberfill. Sew the opening in the bottom.

Sew two of the leg pieces, right sides together. Turn the leg right side out, stuff then pin the foot pad into place. Whip stitch the pad onto the leg opening. Repeat for the second leg.

Sew two of the arm pieces, right sides together. Leave an opening for stuffing. Turn the arm right side out, stuff, then whip stitch the opening closed. Repeat for the second arm.

To attach the arms, mark the spot where the arms are to be attached. Using a long needle and nylon or upholstery thread, go through the right arm, into the body, then out the other side. Knot the thread firmly. Attach the left arm the same way; just be sure not to go through the right arm.

Attach the legs in the same manner.

If the bear is to have movement, add a button to the outside of the arm and leg joints and push the needle through the buttons to secure them. The buttons will strengthen the stitching and make the bear poseable. To finish, stitch around the nose, as designated on the pattern, to make a snout. Use short stitches

It's hard to believe this bear is in his second life. He began as a mohair/cashmere blend coat from the 1940s. Due to the durability of this fabric, vintage coats can still be found, often for a reasonable price. Some have fur collars, unique buttons, and satin linings—all of which are materials for future projects. I like using things that had a previous identity. The creative eye always looks for a new use for old things.

The back of the bear with the button detail. The buttons allow for some movement of the limbs and are reminiscent of an antique bear. Add vintage or colored buttons to coordinate with the fabric. This bear is for big girls and boys only, due to the choking hazard of small buttons.

and then pull taut to gather in the area. This will add definition. Take a small piece of black wool, roll it into a ball, and needle felt it into place for the nose. Add two stitches (using heavy black thread) to form the mouth. For the eyes, using your upholstery needle, run a piece of heavy thread from one side of the head to the other, creating definition for the eye socket. Add a small ball of wool and

needle felt into place to form the eye. Repeat for the other eye. For pads on the feet, add a medium size ball of black wool and needle felt into place for the center. Add smaller balls and form the toes. Needle felt into place. The red heart is made with two small balls of red wool needle felted into place and drawn out to form the heart. Finish the bear with your own hand-felted hat or accessories.

Teddy Bear

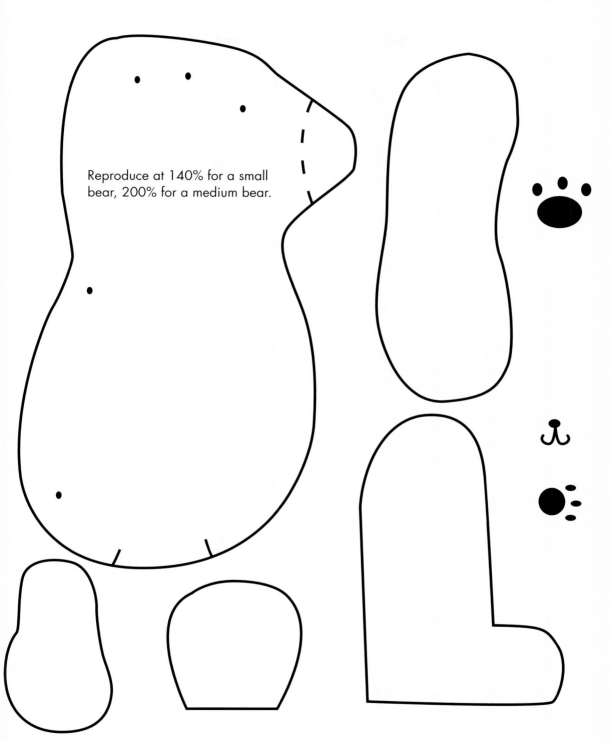

Reproduce at 140% for a small
bear, 200% for a medium bear.

Push-Mold Figures

To try doll-making on a small scale, check out push molds. The molds are incredibly detailed, and with a few additions, such as beads for eyes and mohair for a wig, a distinctive character will emerge! Your own personality will come through. When creating the troll, I laughed out loud when I added his ears. With mohair in place for his toupee, this creature has attitude.

Materials
- Push molds (face, hand, and foot molds)
- Polymer clay (white or flesh colored)
- Water color or acrylic paints to add features
- Seed beads for eyes
- Felt or other fabric for costume
- Muslin for body
- Fiberfill

Before you begin, work the polymer clay to soften it. Begin by chopping the block of clay into tiny pieces. A single-edge razor blade works well for this step. Then add 1 drop of vegetable oil and work the clay for 1 or 2 minutes until it is pliable. Softening the clay will allow it to take the form of the mold without tearing or missing details of the mold.

When the clay is soft, brush the mold with cornstarch and press it into the mold. Remove the clay carefully. Slight changes can be made to the features to personalize the piece. Want a longer nose or pointed ears? Manipulate the clay until the desired effect is achieved. Using a skewer, poke holes so you can add tiny beads for the eyeballs later. These holes will add depth to the eyes, making them more realistic. Press the clay into the hand and foot molds; then, shape them as desired. Remove any excess clay and make small alterations using manicure scissors. Add a hole to the wrists and ankles; you'll place a pipe cleaner in this hole after baking the figure. Bake the clay according to the manufacturer's instructions. When the pieces are cool, paint the features. A little blush for the cheeks and color on the lips and eyes will bring the face to life.

The body is very simple. Cut the body from muslin, sew the pieces together, and stuff it. Using the pattern, make the coat and pants. Felt, corduroy, or a recycled wool fabric works perfectly. Sew the pieces together, wrong side out; then, turn them right side out. Hot glue the head to the body, then put his pants on, followed by his coat. Hot glue the pieces into place. Stuff the sleeves and pant legs lightly. Poke a length of pipe cleaner into the holes in both wrists and ankles, and then work these pieces into the sleeves and pants. Pose the creature as you wish. Stitch the jacket closed, and add interesting charms or embellishments.

Secret Keeper Variation. Follow directions as above for the face and hands. Shape a small bowl out of clay. Bake all pieces. Cut out the small body pattern; sew and stuff it lightly. Add the coat, stuff the sleeves, and add the hands/arms on pipe cleaners. Shape into the desired pose, add the bowl, and glue all of the components in place. Add a few handmade vegetables made of clay for *Woman of the Harvest* or a tiny note for a *Secret Keeper.* (Bits of dyed green wool make great carrot tops!) Use your imagination to add hair, dreadlocks, or whatever else comes to mind to create your own figure. The more off beat, the better!

Troll Variation. Using the directions above, form the troll's face. Add ears of your own design, making them a little pointy for effect. Paint the face and features. For the eyes, add a piece of tapioca for the pupil, centering it in the middle of the eye. Paint the white of the eye, then the iris. Paint the tapioca pupil black. Add a tiny dot of white on top of the black. Add rosy cheeks and lips. Cut, sew, and stuff the body. Hot glue the head to the body. Cut out the troll jacket and pants. Affix the hands and feet to pipe cleaners, and add them to the figure. Then put a small amount of stuffing into the sleeves and pant legs. Add the hair and hat. The hair is what makes this guy! Pose him to guard the bridge. "Trip, trap. Who's that walking on my bridge?"

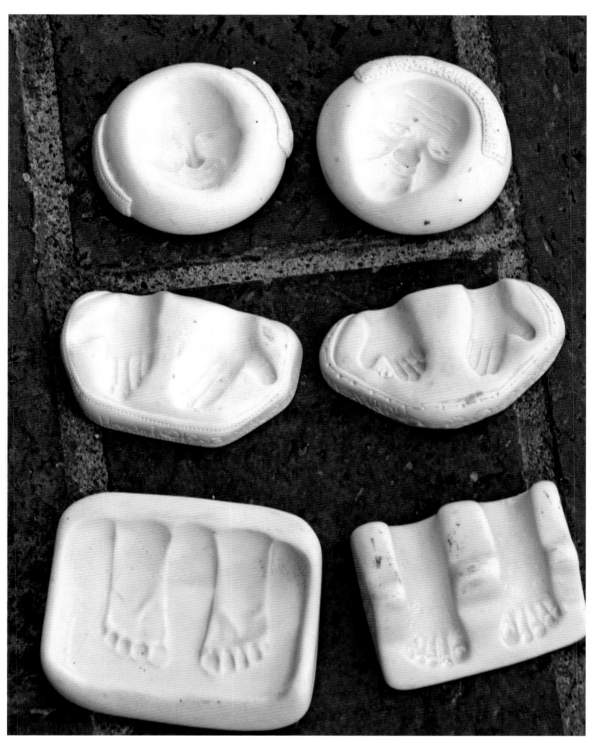

These are push molds for polymer clay. Amazing detail can be captured using these molds. I like to add or change features by stretching the clay, adding more clay to shape an oversized nose, or working with the piece to create an eyelid or open the mouth. Once the basic shape is formed, it is simple to change the basic features. Add a piece bead or ball of tapioca to make the pupil for the eye. Small details add a great detail of impact.

Note the closed eyes and peaceful, knowing look of the Secret Keeper. Her face and hands are made using a push mold. The basic features are altered to give her a distinct personality. Use the pattern provided to shape her cloth body. Add a clay base to stabilize her and a pinch pot clay bowl covered in handmade felt. Add a secret wish, and cover the bowl with a felt-covered lid.

Secret Keeper Push-Mold Figure

Reproduce at 130%

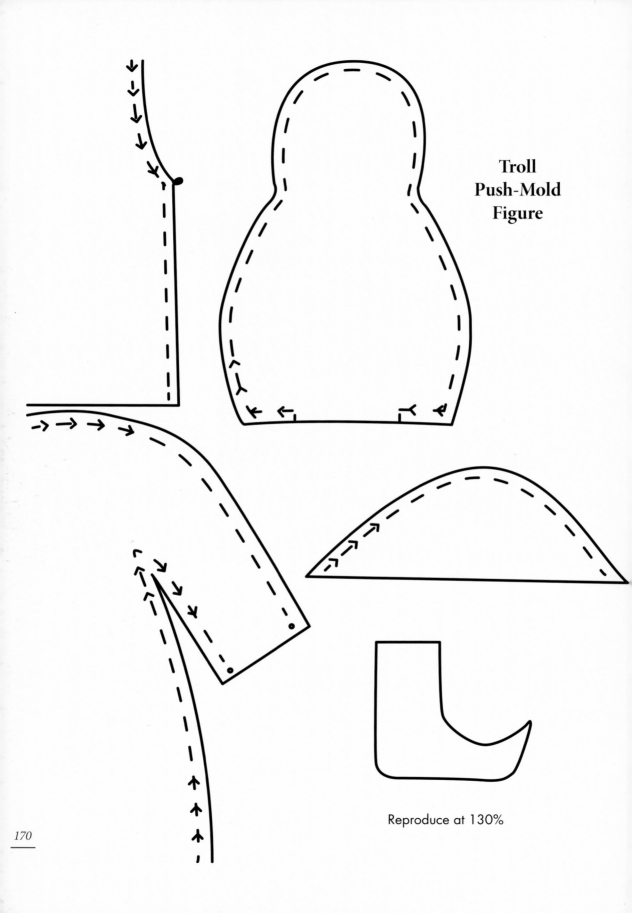

**Troll
Push-Mold
Figure**

Reproduce at 130%

What book on goats would be complete without a troll? Made famous in the "Three Billy Goats Gruff" tale, this fellow is ready to guard the bridge. His face was made of polymer clay in a push mold, then his nose and ears were added to give him his trollish look. He has a clay water bottle fashioned for his journey and a raven fashioned of clay. "Trip, trap. Who's that walking on my bridge?"

If Santa took a walk in the woods, he would undoubtedly collect things of nature along the way. With his sticks and feathers, this fellow is on display year round in my woodland home. His coat is from a thrift-store jacket, and the fur collar is from a vintage coat. A coat of clear nail polish on his eyes gives a secret sparkle to his gaze.

Woodland Santa

What project do I think of when I see mohair? Santa! Just looking at mohair brings a Santa to mind. Whether the jolly old elf is done in traditional red or a woodsy theme for Santa of the Woods, there is no limit to the number of ways to design a Santa with personality plus. For the Santa pictured (opposite), the same face as for the troll was used, and it is remarkable how a change of clothing changes his demeanor. Try a Santa done in deer skin for a woodland theme, an all-white Santa, or how about one with skis? The possibilities are endless. Once I found a pair of tiny wooden clogs that inspired a woodland Santa with a handmade pack basket, greenery, and a tiny lantern. It is a joy to make something that is one of a kind. Santa outfits can be made of felt, wool, corduroy, velvet, or any heavyweight fabric. Collect small toys, bits of greenery, or little antique pieces such as bells or ornaments to adorn Santa's pack.

Merlin is another variation on this theme. The same pattern is used for his coat. His hat is stuffed to make it stand up on top of his head. With his crystal ball (a reproduction drawer pull), he looks ready to make magic!

Materials

• Santa face (This may be purchased premade, hand-sculpted, or cast. This Santa face is from Country in the City [www.countryinthecity. net], my longtime supplier of resin faces.)
• Fabric, for the body (Muslin is fine.)
• Fiberfill
• Felt, for the boots
• Wood, for the stand
• Dowel, for the stand
• Wire, for the stand
• Fabric for the coat, pants, gloves, boots, pack, etc.

Paint the face with flesh-colored acrylic craft paint, let it dry, then attach the face to the muslin with hot glue. Paint the eyebrows using white paint. Move to the eyes. The eyes are the details that add life and make the face authentic. Paint the eye opening white. Let it dry. Using blue paint, add a small circle to the center of the white eye opening for the iris of the eye. Let it dry. Next add a smaller black circle inside the iris for the pupil. Last, add a tiny drop of white in the right-hand corner of the pupil. This drop makes the eye sparkle a bit. Now paint the lips a cherry red. Thin out the red paint with water and gently brush the cheeks to give them a blush. If an antique look is desired, thin out some brown paint and do a light wash over the entire face. Remove some of the wash with a damp paper towel. Allow the wash to remain in the nooks and crannies of the face, eyebrows, and eyes to give the face a warmer tone. Let the face dry.

Cut out the body from muslin. Sew the body together, allowing for a ⅝-inch seam allowance. Leave a 2-inch opening at the bottom to add stuffing. Stuff the body firmly with fiberfill, and whip stitch the opening closed.

Cut out the coat, and sew the pieces together with a ¼-inch seam allowance. Sew with the finished side of the fabric to the inside. Turn the fabric right side out, and cut an opening down the front of the coat.

Cut out the pants; sew with ¼-inch seam allowance with the right side of the fabric turned in. Sew one leg, then the other, then join them with a center seam (see pattern, page 175). Place the body in the finished pants, and add the coat.

Glue the face to the body using hot glue. The Santa can be either sitting or stand-

continued on page 177

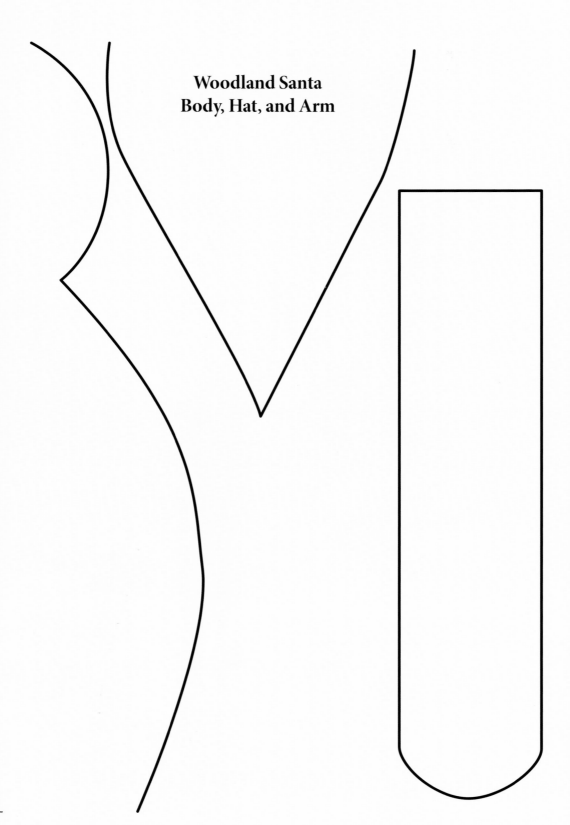

**Woodland Santa
Body, Hat, and Arm**

Woodland Santa
Legs and Boots

Reproduce both at 160%

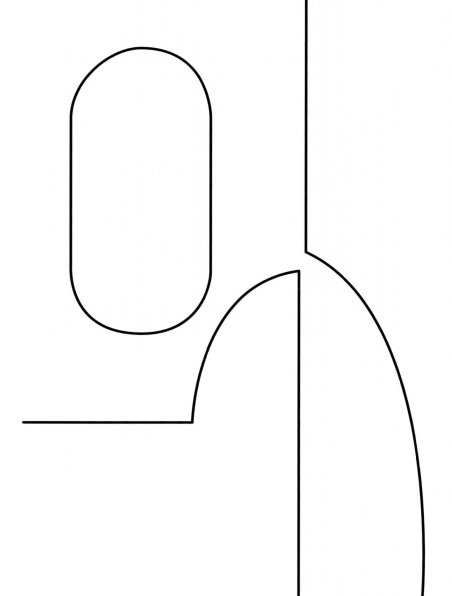

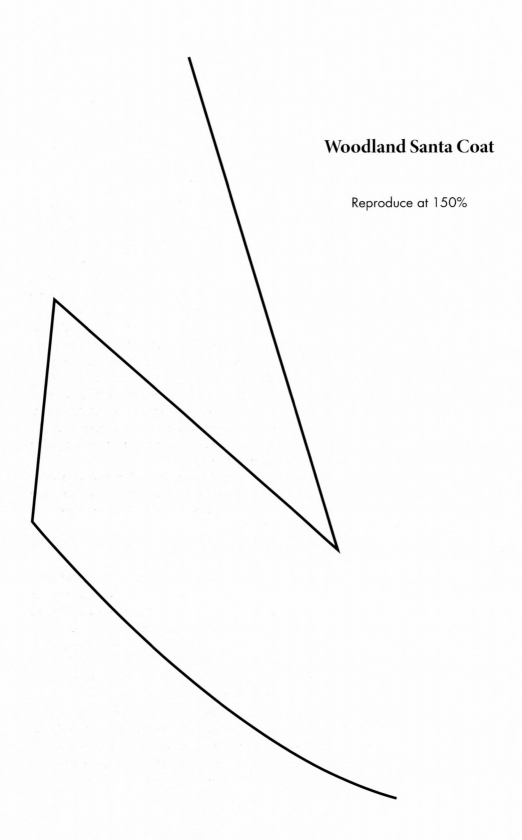

Woodland Santa Coat

Reproduce at 150%

continued from page 172

ing. To stand him up, purchase or make a doll stand and affix him to the stand. For a sitting Santa, stuff the legs of the pants with fiberfill.

Stuff the jacket sleeves with fiberfill to create arms. Add molded hands made with a push mold and polymer clay. Mittens made of felt or tiny knitted mittens are fine, too. Hot glue the hands or mittens in place, gathering the sleeve around them and using glue to affix the hand to the sleeve.

Add locks of mohair to create the beard. Place a lock at a time, hot gluing each in a sequence starting at the bottom of the face and layering upward until the desired fullness is attained. Twist a small piece of mohair for the mustache.

For the hair, begin adding mohair locks at the back of the head at the bottom. Add a row of mohair at a time, building the hair so it has a natural appearance. Continue to layer the locks until the head is covered up to the solid part of the face. Add a little curl across the forehead and another on each side of the molded face so it appears that the hair is surrounding the face in a natural growth pattern (see the finished example on page 173).

Add trim to the coat and purchased or handmade felt down the center and around the edges of the coat. Cut the felt in ½-inch strips and hot glue it onto the coat fabric. Add felt at the junction of the hands or mittens to finish the sleeves. Do the same at the base of the pants to finished off the "hem" at the boot and pant leg.

Sew a pack, if desired, out of a contrasting fabric, and fill it with small toys, a bit of greenery, and other items. Wait to attach the pack until after Santa is on his stand.

Santa can sit in a doll's chair (look for those at yard sales), or he can stand. To make a simple stand, cut or purchase a 6-inch wooden circle of pine or purchase a pine plaque at a craft store. Drill a ¼-inch hole 1 inch from the edge of the circle. Sand the entire piece. Cut a length of ¼-inch dowel rod 12 inches in length. Add a drop of wood glue to the hole, then stick the dowel into the hole and let it dry. Paint the entire piece black. Cut a piece of lightweight wire, aluminum, copper, or other type wire that is sturdy but flexible. The length should be 6 inches. Wrap the wire one time around the dowel just about three-quarters of the way up. Add a little hot glue to keep it firmly in place. When the glue is set, pick up the Santa and pull up his jacket. Place him on the stand, and encircle his belly with the wire to keep him upright. Rearrange his jacket to hide the wire.

Add the belt. Use velvet ribbon, felt, or leather. Wrap it around the waist, and add a pretty button or something to appear as a buckle.

Now bring on the pack, toys, or whatever accessories fit the theme.

Let your imagination run wild with your Santas. I have made artist Santas with an easel and palette, woodsmen, Hawaiian Santas, fisherman Santas, Santas with a cello—all sorts of new takes on a familiar theme. Just recently I saw an Americana Santa with a blue jacket covered in stars and red striped pants. That's my next project!

I look for toys and other items scaled to Santa size and keep those on hand to use in my creations. For several years, making Santas and dolls was a very nice cottage industry for my family. No one can resist a Santa!

Goat-Milk Soaps and Lotions

Goat-milk soap is a natural item to add to the goat product repertoire. Soap making is an ancient art that has come back into the forefront of handmade goods. The qualities of the goat milk make it a natural moisturizer, very soothing to the skin and thus a natural addition to the soap maker's pantry. The use of goat milk in the soapmaking process adds a natural emollient that is beneficial to those with sensitive or dry skin. When you make your own soap, you control the quality of ingredients, the amount of scent added (if any), and the use of colorants or other additives. If you have allergies or sensitivities to chemical additives found in many commercially made soaps, try making your own blends.

In the soapmaking kitchen at Double Diamond D Dairy. A wooden soap mold handmade by Dale Deweese and the Fresh Milk sign create the perfect still life.

It is very important to weigh the ingredients for soapmaking. Don't rely on a measuring cup. Exact measurements are very important for overall success when making soap.

This first recipe is one that I make on a frequent basis. It calls for lard. While some may recoil at the use of an animal product, traditional soap making always included lard. This soap is unscented. Expect goat-milk soap to turn a rich honey brown color.

An Important Word About Lye

Some soap recipes call for lye, which can be very dangerous to work with. I once made a mistake and used an aluminum pan with lye. I created an extremely dangerous situation when a gas formed and filled the house. **Do not use aluminum pans when making soap!** Also, use goggles, long sleeves, and rubber gloves. Don't make soap without wearing shoes. If the lye gets on your skin, it is going to burn! Finally, keep children out of the soapmaking kitchen. This is an adults-only project.

Traditional Goat-Milk Soap

Equipment

Scale, Glass measuring cup, Wooden spoon
Enamel pot (Designate this pot for soap making and do not use it for food preparation in the future.)
Stainless steel bowl
Hand mixer or old hand crank rotary blender
Stick blender (recommended but not essential)
Petroleum jelly
Thermometer
Cardboard flat, such as canned goods come in, or a wooden soap mold
Old towels

Ingredients

3 cups goat milk
½ pound lard
6.5 ounces lye (Lye is difficult to find on a store shelf. Order lye from a chemical company. See the resource list, page 187.)
2 ounces glycerin

1. If using a cardboard box, line it with a plastic bag, fully covering the cardboard. Make sure the ends of the box are not open or the liquid soap will run out. If using a wooden soap mold, grease the mold with petroleum jelly. Do not use a plastic soap mold or a candy mold. This soap will be hot when it is poured into the mold, and the plastic will melt.
2. Place the lard in the soap pot and melt, using a low heat. After melting, cool the lard to 90° F.
3. Place a paper cup on a scale, and measure out 6.5 ounces of lye. Pour 3 cups of cold goat milk into a stainless steel bowl, and slowly add the lye. Mix this up with the hand mixer or, my

preference, an old rotary egg beater. Mixing the milk and lye will create a chemical reaction and the milk will turn a honey brown. Cool the mixture to 85° F. Add the glycerin. Using the electric hand mixer or, even better, a stick blender to expedite the process, mix/blend the mixture until the liquid starts to change consistency. This can take up to 30 minutes! Mix a bit, then wait a few minutes and mix again. All of the sudden, there will be a drastic change in consistency. This is called the "trace." Check for the trace by bringing some of the soap in progress onto the wooden spoon. Move the spoon above the soap in the pot to form a line of liquid on top of the mixture. If the drizzle stays on top for a few seconds before going back into the mixture, the trace stage has been reached. If not, blend some more and test

Bars of soap drying in the soap maker's kitchen for a few days after cutting. The curing process takes 6 weeks. Goat milk is a natural emollient and adds moisture to the skin. Deb freezes the milk for soap.

again. When the trace stage has been reached, stop blending. Pour the soap into a prepared mold (cardboard or wooden), and cover the hot soap with wax paper. Wrap the box in a towel, then add more towels or blankets to keep the heat in as long as possible. Keep the box of soap in a warm place (room temperature) for 24 hours.

4. The next day, remove the towels and check the soap; it should be firm. If it isn't, leave it in the mold another day or two. Cut the firm soap into uniform bars with a cheese slicer, a wire, or a knife. Place the bars on a clean piece of cardboard or on a wooden tray and air dry them for about 6 weeks. During this time, the soap will harden and cure. The lye remains harsh until after this curing period.

5. This tried and true recipe comes from Jeanette Larson, a long time goatkeeper. Jeanette has had a herd of dairy goats since 1979 and about 10 years ago added fiber goats. She states, "Goats are a vital part of my life." This is a common statement by those who have or are owned by goats! Follow Jen on her blog at www. jenonthefarm.com.

Deb DeWeese uses a scale for accurate measurement while making soap. Soapmaking is a frequent activity at the Double Diamond D Dairy in Lonedell, Missouri.

Lavender Goat-Milk Soap

Ingredients

6 pounds generic soy shortening

40 ounces canola oil

24 ounces lye

2 ounces safflower oil

9 ounces olive oil

2 ounces grape seed oil

6 pounds frozen goats milk (Jeanette states, "Freeze it in 2-pound chunks in plastic freezer bags.")

½ cup lavender flowers

50 drops lavender essential oil

1. In the very biggest stainless steel pot, melt the shortening and blend it with the oils on low heat. Stir and check the temperature frequently to keep the mixture below 120° F.

2. In a separate, large stainless steel pot (both pots should be large enough to hold the entire batch), place the slightly thawed, still very slushy goat milk. Stirring constantly, pour the lye very slowly; (take at least 10 minutes for this step!) over the goat-milk slush and blend it in with a mixer or stick blender. It will turn a creamy lemon yellow and should stay under 120° F.

3. When the oils are all blended in one pot and the liquid and goat milk are smoothly blended together in the other pot, pour the milk mixture into the oils, stirring as you do so. From here on out, you need to stir constantly, using your mixer or stick blender.

4. Jeanette states, "My set up for the next stage is a big, strong, linoleum-covered table. I put one of the pots on one side and the other one on the other side and a heavy-duty blender in between. Using a ladle, I scoop the soap mixture into the blender, filling in about half full. I run the blender for about a minute and pour the blended soap mixture into the second pot. You must keep stirring both pots. I repeat this, going back and forth through the blender from pot to pot. When the soap is starting to trace, I add the lavender flowers and essential oil, dividing it out through about three blender loads. This soap usually goes through the blender twice before tracing, but occasionally it traces after only one pass through the blender—often it takes three passes through, depending upon the day."

5. At trace, pour the soap mixture into plastic molds, 15 x 2 x 3½ inches. This batch will fill seven or eight of these molds. Cover the filled molds loosely with paper towels to keep dust out. When they are firm, put them in the freezer for 4 to 6 hours and then pop them out of the mold. You can cut it then, but I usually let it cure 6 weeks in the block and then cut it right before wrapping.

6. Wrap the cut soap in paper or cloth so the soap can continue to breathe. It is very high in oils, which will leach out if the soap is wrapped in plastic.

Freshly made goat milk soap poured into a mold. The soap will sit, covered, overnight and be ready for cutting the following day.

Deb cuts soap. The soap cutter adds a nice ridge design. The molds are made on the farm by Deb's husband, Dale. Deb makes soap several times a month and is a favorite vendor at the Route 66 Famers' Market.

Soap in a Loofah

Loofah sponges are great for the shower. A natural scrubber, filled with soap makes them even better.

Have the loofah ready to receive the hot soap before you begin. Wrap the sponge in plastic wrap, and secure the ends of the wrap with tape. Make sure the sponge is wrapped securely or the soap will run out. Place the wrapped loofah, upright, in a box.

To make soap in a loofah, use any soap recipe. When the soap reaches the trace stage pour the semiliquid soap into the cavity in the loofah. Place the filled loofah in the freezer, making sure it is still sitting upright. Remove the loofa after several hours, peel off the plastic wrap, and let the soap cure for 6 weeks. Slice the loofa like a jelly roll so the inner part of the natural sponge is visible.

A note about oils: For my soaps, I always use essential oils rather than fragrance oils. My goal in making soap is to use as many natural ingredients as possible. Essential oils are distilled from the plant. Fragrance oils are manmade. I also prefer not to color my soaps for the same reason. For me, the fewer colors and manmade products added, the better my soap will be.

Goat-Milk Bath Tea

This simple, four-ingredient recipe from the collection of Anne-Marie, the Soap Queen at Brambleberry Soap Making Supplies only has four steps, making it easy to whip together a luxurious bath after a hard day on your feet. What's so luxurious about goat milk? Goat milk is particularly **moisturizing and nourishing** to the skin because of capric-caprylic triglyceride. Capric-caprylic triglyceride is an effective skin moisturizer that helps contribute to skin softness by forming a barrier on the skin that helps to inhibit the loss of moisture. It is the only milk that contains naturally occurring capric-caprylic triglycerides. The protein strands of goat milk are shorter than in other types of milk and are more readily absorbed by skin.

Goat milk also has naturally occurring lactic acid that **helps to keep skin smooth** by encouraging skin turnover (it acts similar to a gentle peel). It also contains many vitamins, specifically A, D, and B6, as well as the antioxidant selenium.

Ingredients

4 tablespoons powdered goat milk
3 tablespoons lavender buds, grade 2
2 tablespoons medium pink sea salt
2 tablespoons cocoa butter (shaved and unmelted)
1 heat sealable large tea bag
Vegetable peeler or Brambleberry Clean Up Tool
Yields 1 tea bag for the bath

1. Shave 2 tablespoons of cocoa butter using a Clean Up Tool (Brambleberry) or a vegetable peeler. Combine all of the ingredients in a medium-sized container, and gently mix them together with a spoon.

2. Carefully scoop the ingredients into the large tea bag. To make a little more room at the top, tap the tea bag on the counter to condense the ingredients. Fill the bag up with all of the ingredients.

3. Carefully tip the tea bag on its side and seal the open side with a hot iron. It's ready for the tub!

When you're ready for a skin-loving soak, just add the tea bag to your hot bath. Even though we didn't add any fragrance oil, it smells creamy and delicious. The cocoa butter melts in the warm water adding a little richness to the goat milk and lavender. I totally love the smell of goat milk.

4. *Tips for the Tub.* The goat-milk powder is super fine, so give the tea bag a couple of big squeezes while submerged in the water to get the goat milk out, or use a pin to poke small holes in the tea bag. Make the holes just big enough to get the goats milk powder out while keeping the salt and lavender buds in.

Want to make more for yourself, to sell, or give as gifts? Just double, triple, or quadruple the recipe as needed!

Goat-Milk Lotion

This one is too easy.

Ingredients
1 cup of goat milk, room temperature
1/8 cup mild olive oil
1/8 teaspoon white vinegar
Add 4 drops of essential oil

Put goat milk and olive oil in a glass mixing bowl, and mix thoroughly. An old-style rotary egg beater is good for this task. When well blended, add the essential oil, then the white vinegar to preserve the lotion. Add a few drops of essential oil, and blend it in. A pump-style container makes a good dispenser for this type of lotion. Shake before using; then refrigerate. Try spearmint essential oil for a summer refresher!

Goat-Milk Lotion with Healing Oils

Another goat-milk lotion recipe, this one is a little more complicated but well worth the effort.

Ingredients
1 1/2 tablespoon beeswax
2 tablespoons olive oil (mild)
3 tablespoons sweet almond oil
2 tablespoons calendula oil
1 tablespoon apricot oil
1 cup goat milk, heated to 85° F
8 capsules of vitamin E oil
Few drops of essential oil of your choice

Combine the beeswax and olive oil in a small saucepan. Heat until the beeswax is melted. Whisk the oil and beeswax until they are blended. Remove the mixture from the saucepan, and place it in a glass bowl. Add the sweet almond oil, the calendula oil, and the apricot oil to the wax mixture. Add the goat milk; then, break open the vitamin E capsules and drizzle the vitamin E oil into the mix. Add the essential oil if desired. Using a mixer, mix until all ingredients are mixed in and the mixture starts to thicken. Let the lotion cool until it reaches room temperature. Store in a plastic container with a lid.

Goat-Milk Soap

This recipe is from the collection of Deb Deweese. Deb and her husband, Dale, operate Double Diamond D Dairy (www.doublediamondddairy.com) in central Missouri. Their herd of 22 dairy goats keeps them busy on the farm. Deb makes cheese on a regular basis and saves some milk out for her soap making. She is a frequent vendor at area farmers' markets and offers her lovely soaps for sale at the market or through her website. Her talented husband, Dale, creates handcrafted wooden soap molds, also available through the website. Registered breeding stock is also available.

Ingredients
5 ounces canola oil
5 ounces shea butter
5 ounces castor oil
20 ounces lard
12 ounces coconut oil
10 ounces goat milk
8 ounces olive oil
11 ounces water
5 ounces palm oil
8.3 ounces lye (sodium hydroxide)
Essential oil (General rule of thumb is 1½ to 2 ounces per 30 ounces of oil)
Note: This recipe has a total weight of oils of 60 ounces.

Equipment
Heavy plastic pitcher, to mix milk, water, lye
One 2-gallon bucket, to mix oils.
Stainless steel, long-handled spoon to mix ingredients
Another bucket, with ice water to set the pitcher in
Electric stick blender
Scale, that measures in ounces
Molds, for your soap
Protective gloves, safety glasses, and long sleeve shirt

Warning: Always add solid form lye (sodium hydroxide or potassium hydroxide) to the liquid. If liquid is added to solid form lye, a violent reaction can result. Remember: Lye is caustic!

Freeze the goat milk in ice cube trays ahead of your starting time. You want to start off with the coldest milk possible to prevent the lye from burning it. Have all of the ingredients ready.

Set the pitcher in the bucket of ice water. Add the frozen goat milk and water to the pitcher. Carefully add the lye. Stir carefully, until all the frozen milk is melted.

Measure the liquid oils, and put them in the mixing bucket. Measure the butter, lard, and coconut oil, and heat them until just about melted in the microwave. Mix them in with rest of the oils. Carefully add the milk and lye solution to the oils. Stir constantly if you are stirring by hand; if using the stick blender, blend for 2 minute intervals.

When the mixture comes to trace, add the fragrance oil and pour it into the molds. Leave the mixture in the molds overnight. Remove them from molds, and age for at least 4 weeks.

Felted Soap

What? Felted soap? This combines two of my favorite things—and I hope yours, too. Take a bar of your handmade soap and wrap it in carded wool and mohair blended together. Wrap it thoroughly so it is completely covered in about a ½-inch thickness of fiber. Slip the wrapped bar down into a nylon stocking. Wet stocking with hot water and work it back and forth with your hands, a soft brush, or on an old wash board. An alternative to a washboard is a colander turned upside down. This provides some friction to work the wool back and forth, creating the felt. The soap will lather up some while you are felting; that's ok. It will assist in the felting process. When the felt is solid, squeeze out any excess water and blot the felt with a paper towel. Allow the felted soap to air dry.

Resources

Books on Goatkeeping
Amundson, Carol A. *How to Raise Goats*. Minneapolis, MN: Voyageur Press, 2009.

Belanger, Jerry and Thomson Bredesen, Sara. *Storey's Guide to Raising Dairy Goats: Breeds, Care, Dairying, Marketing*. North Adams, MA: Storey Publishing, 2010.

Drummond, Susan. *Angora Goats the Northern Way, 5th Ed.* Freeport, MI: Stoney Lonesome Farm, 2005.

Zweede-Tucker, Yvonne. *The Meat Goat Handbook: Raising Goats for Food, Profit, and Fun*. Minneapolis, MN: Voyageur Press, 2012.

Books on Cheesemaking
Carroll, Ricki. *Home Cheese Making: Recipes for 75 Homemade Cheeses, 3rd Ed.* North Adams, MA: Storey Publishing, 2002.

Farnham, Joy and Druart, Marc. *The Joy of Cheesemaking*. New York: Skyhorse Publishing, 2011.

Hurst, Janet. *Homemade Cheese: Recipes for 50 Cheeses from Artisan Cheesemakers*. Minneapolis, MN: Voyageur Press, 2011.

Kindstedt, Paul. *American Farmstead Cheese: The Complete Guide to Making and Selling Artisan Cheese*. White River Junction, VT: Chelsea Green, 2005.

Toth, Mary Jane. *Goats Produce Too! The Udder Real Thing, Vol. 2: Cheesemaking and More, 6th Ed.* self-published, 1998.

Books on Felting and Weaving
Hoerner, Nancy et al. *Easy Needle Felting*. New York: Sterling, 2008.

Horvath, Marie-Noelle. *Little Felted Animals: Create 16 Irresistible Creatures with Simple Needle-Felting Techniques*. New York: Potter Craft, 2008.

Matthiessen, Barbara. *Small Loom & Freeform Weaving: 5 Ways to Weave*. Minneapolis, MN: Creative Publishing international, 2008.

Rainey, Sarita R. *Weaving Without a Loom, 2nd Ed.* Worcester, MA: Davis, 2008.

Warner Dendel, Esther. *Needleweaving . . . Easy as Embroidery*. New York: Doubleday, 1976.

Cheesemaking Supplies
Hoegger Goat Supply, www.hoeggerfarmyard.com

New England Cheesemaking Supply Company, www.cheesemaking.com

Soapmaking Supplies
Bramble Berry, www.brambleberry.com

Technical Resources
ATTRA, www.attra.org – National Sustainable Agriculture Information Service

CheezSorce, www.cheezsorce.com – Neville McNaughton and his team of cheesemaking consultants can help you set up your facility with the right equipment and systems in place. Along with their expert advice in the dairy and cheese industry, they travel the world organizing and promoting cheesemaking classes and workshops.

David Fisher, www.candleandsoap.about.com – The online guru for candle and soap makers

SARE, www.sare.org – Sustainable Agriculture Research and Information

University of Minnesota Extension, www.extension.org – Research-based information from the nation's oldest and largest network of university experts

Research-Based Goat Programs

Cornell University, www.cornell.edu

Langston University, www.langston.edu

Lincoln University Cooperative Extension (LUCE) and the Innovative Small Farmer's Outreach Program (ISFOP), www. lincolnu.edu

Texas A&M, www.tamu.edu

Cheesemaking Instruction and Education

The Vermont Institute for Artisan Cheese at the University of Vermont (www.nutrition.uvm. edu/viac) is the nation's first and only comprehensive center devoted to artisan cheese. By providing education, research, technical services, and public service to increase knowledge, appreciation, and expansion of artisan cheese, the institute supports artisan cheese producers throughout the United States, contributes to the latest scientific research and expertise related to dairy and cheese products, and encourages the sustainability of the small-farm culture in Vermont and other rural areas.

Magazines

Dairy Goat Journal, www.dairygoatjournal.com – A wonderful resource for the goatkeeper.

MaryJanes Farm, www.maryjanesfarm.com – Graceful and lovely, MaryJane Butters is a farmgirl from the heart.

Mother Earth News, www.motherearthnews.com – What can I say about *Mother*? Reading this magazine got me started down the goat path all those years ago.

Small Farm Today, www.smallfarmtoday.com – Their tagline says it all: The Original How-to Magazine of Alternative and Traditional Crops and Livestock, Direct Marketing, and Rural Living

Small Farmers' Journal, www.smallfarmersjournal.com – Great features on draft animals and stories of farm life.

Goat Cheesemakers Extraordinaire

Veronica and Steve Baetje, www.baetjefarms.com

Jennifer Bice, Redwood Hill Farm, www.redwoodhillfarm.com

Doe's Leap, www.doesleap.com

Mary Keehne, Cypress Grove Chevre, www.cypressgrovechevre.com

Lazy Lady Farm, www.lazyladyfarm.com

Jenn and Ken Muno, www.goatsbeardfarm.com

Vermont Butter & Cheese Creamery, www.vermontcreamery.com

Goatkeeper's Supplies

Caprine Supply, www.caprinesupply.com

Hoegger Goat Supply, www.hoeggerfarmyard.com

Lehman's, www.lehmans.com

Organizations

American Angora Goat Breeders Association, www.aagba.org

American Boer Goat Association, www.abga.org

American Cheese Society, www.cheesesociety.org

American Dairy Goat Association, www.adga.org

Many areas have goat clubs and weavers' and cheesemakers' guilds. Check online listings.

Acknowledgments

Many thanks to Michael Dregni and Kim Carr, who bring my words to life; to sister-friend Joan Treis of the Hermann Farm (www.hermannfarm.org) for providing moral support and the excellent venue for photography. Hermann Farm is a 160-plus-acre working farm, living-history center, and National Landmark located along the Missouri River, about an hour west of St. Louis at the entrance to scenic Hermann, Missouri.

It's hard to believe our time has come and gone. I hope you have been inspired and will try some new projects, maybe even add a goat to your family. Whatever direction life takes you, good wishes follow you from my farm to yours. Blessed are the Goatkeepers.

Here are a few words of wisdom I learned from a goat:

Beware of trolls.

Horns happen.

Why walk when you can scamper?

Good things come in pairs.

Eat out at least once a week.

Fences are merely suggestions.

Never underestimate a farm girl.

Naps in the sunshine are good for you.

Joy is contagious.

Aim for the high places.

Index

Items for which a recipe is provided appear in bold.
Page numbers in italics indicate a photograph.